Sunset

Garden
Pests & Diseases

By the Editors of Sunset Books and Sunset Magazine

Ladybird beetles, also known as ladybugs, help gardeners by preying on soft-bodied pests. This convergent lady beetle is about to devour a potato aphid.

Sunset Publishing Corporation ▪ Menlo Park, California

From its vantage point atop a tomato, a baby American toad hunts for insect prey.

Book Editor
Lynne Gilberg

Research & Text
Susan Lang

Coordinating Editor
Paula Goldstein

Design
Joe di Chiarro

Illustrations
Lois Lovejoy

Cover: Tarnished plant bug on a lima bean leaf. This pest sucks juices from plants, at the same time injecting a toxin which deforms the foliage. Cover design by Susan Bryant. Photography by Runk/Schoenberger/Grant Heilman Photography Inc.

Editor, Sunset Books: Elizabeth L. Hogan
First printing February 1993

MANAGING PESTS THE IPM WAY

Those of us who began gardening after the pesticide revolution of the 1940s were led to believe that chemicals were miracle weapons—but this myth has been shattered by the revelations of the past few decades. Some of the "wonder" chemicals caused environmental damage, while others proved to be carcinogens; and in many cases, the target pests became resistant to the control.

Today, many commercial growers have freed themselves from dependence on pesticides and opted for *integrated pest management (IPM)*. This approach focuses on prevention and advocates physical and biological controls first, followed by (minimal) chemical action *only* when all the evidence is in.

Symbolic of the trend toward IPM is a change in the address of the USDA Research Center in Beltsville, Maryland: in 1990, the Center renamed its street, abandoning Pesticide Road in favor of Biocontrol Road.

Now IPM has moved from commercial fields into home gardens. With the help of this book, you'll learn everything you need to know about IPM principles and techniques. It's time to shelve those pesticide bottles for the moment and start working on a new kind of pest management program.

For reviewing the manuscript, we extend very special thanks to John R. Dunmire, and to Richard Cowles, Arthur McCain, and Rex Marsh of University of California. For their assistance in providing information and ferreting out little-known facts, we wish to thank the USDA and the Cooperative Extension Services of Cornell University, University of California, University of Illinois, and University of Massachusetts.

Finally, we gratefully acknowledge Rebecca LaBrum for her judicious and careful editing of the manuscript.

Photographers

Scott Atkinson: 40, 45, 46; **Max E. Badgley:** 53 left, 79 top; **Liz Ball/Photo/Nats:** 74 top; **Thomas C. Boyden:** 6, 20, 22, 34, 63, 105 middle; **Larry Brock/TOM STACK & ASSOCIATES:** 16 left; **Gay Bumgarner/Photo/Nats:** 2, 70, 93 bottom, 95 bottom; **California Department of Forestry and Fire Prevention Association:** 100 middle; **Scott Camazine:** 38 bottom middle, 72 bottom, 78 middle, 79 bottom, 81 top, 82 top, 90 middle, 91 top and middle; **James L. Castner:** 4 right, 15, 37 top left, 72 top, 74 middle, 76 bottom, 77 top, 81 bottom, 83 bottom, 85 bottom; **Peter Christiansen:** 99 top; **Jack K. Clark/COMSTOCK, INC.:** 1, 52, 53 right, 55, 80 top, 83 top, 84 top, 85 middle, 86, 87 top, 103 bottom left; **Priscilla Connell/Photo/Nats:** 76 bottom, 87 bottom; **W. Perry Conway/TOM STACK & ASSOCIATES:** 85 top; **John Cooke/COMSTOCK, INC.:** 38 bottom right; **Richard S. Cowles:** 88 bottom; **Crandall & Crandall:** 4 left, 24, 26, 35, 44, 49 middle, 57, 59, 60, 102 bottom left; **Greg Crisci/Photo/Nats:** 91 bottom; **Derek Fell:** 49 left and right, 50, 73 bottom, 79 middle, 98, 99 bottom, 100 top, 101 top, 104 upper middle; **Kerry T. Givens/TOM STACK & ASSOCIATES:** 102 top left; **Grant Heilman/Grant Heilman Photography Inc.:** 73 middle, 78 top, 87 middle; **Valerie Hodgson/Photo/Nats:** 93 top; **Bill Ivy:** 38 top right, 94 bottom, 97 bottom; **Dr. Jerral Johnson:** 104 bottom, 105 top; **George D. Lepp/COMSTOCK, INC.:** 75 bottom, 88 middle, 95 top, 96 bottom; **John A. Lynch/Photo/Nats:** 10 right; **Robert E. Lyons/Photo/Nats:** 92 top; **Stephen G. Maka/Photo/Nats:** 39 top right; **Jeff March/Photo/Nats:** 38 bottom left, 89 top; **Joe & Carol McDonald/TOM STACK & ASSOCIATES:** 97 top; **Sam McVicker/COMSTOCK, INC.:** 94 top; **Robert & Linda Mitchell:** 5, 9, 16 right, 36 right, 37 top middle, 38 top left, 74 bottom, 77 bottom, 78 bottom, 80 middle, 81 middle, 89 middle, 90 top and bottom, 103 middle left; **Richard Molinar:** 99 upper middle, 100 bottom, 104 lower middle; **Mycogen:** 68; **Brad Nelson/Nelson-Bohart & Associates:** 37 bottom middle, 39 middle, 84 middle; **Arleen Olson:** 39 bottom right, 88 top; **Muriel Orans:** 27, 61; **Herbert B. Parsons/Photo/Nats:** 38 top middle; **Paul Peterson:** 37 bottom left; **Rod Planck/TOM STACK & ASSOCIATES:** 83 middle; **Norm Plate:** 28; **Robert D. Raabe:** 99 lower middle, 105 bottom; **James H. Robinson:** 36 left and middle, 37 right, 82 middle and bottom; **Runk/Schoenberger/Grant Heilman Photography Inc.:** 73 top, 75 top, 76 middle, 80 bottom, 84 bottom, 92 bottom; **The Scotts Company:** 102 middle and right, 103 top left and top middle; **John Shaw:** 39 top and bottom left; **Richard Smiley:** 103 top right; **Donald Specker:** 56, 75 middle; **David M. Stone/Photo/Nats:** 12; **Michael S. Thompson/COMSTOCK, INC.:** 10 left and middle, 14 bottom, 31, 42, 43, 48, 54; **Darrow M. Watt:** 104 top; **D. Wilder/TOM STACK & ASSOCIATES:** 77 middle; **Muriel V. Williams/Photo/Nats:** 89 bottom, 96 top; **Thomas A. Zitter:** 14 top, 101 middle and bottom.

CONTENTS

SPECIAL FEATURES

A NEW APPROACH TO PEST MANAGEMENT

No garden, not even a well-tended one, is immune to pests and diseases. Despite your best efforts to banish them, destructive organisms will fly, crawl, creep, leap, or amble into your garden from neighboring properties. Some will even drift in on the wind: air currents can carry certain pests, such as gypsy moth caterpillars, for miles.

Until recently, the typical home gardener—following the example of most farmers and other commercial growers—depended on chemicals to get rid of garden troublemakers. The introduction of "miracle" pesticides like DDT in the 1940s engendered a long-standing conviction that laboratory-produced chemicals were the solution to all pest problems; for decades, the prevailing philosophy held that the only good bug was a dead bug.

The excitement over synthetics isn't difficult to understand. These controls were cheap, easy to use, and seemingly effective. They produced immediate, deadly results, causing pest-control practitioners to jettison other techniques that worked more slowly or produced a less impressive body count. Some time-honored tactics, such as sifting ordinary dust from a dirt road and sprinkling it on plants to discourage insects, now seemed ridiculously "low-tech." Even the practice of pitting beneficial insects against destructive ones, begun in the United States in 1889, was swept aside in favor of the powerful new pesticides.

Large, slow-moving pests, such as hornworms, are easy to handpick. If you're squeamish, use kitchen tongs to grab the pests and drop them into a jar of soapy or oily water.

This alligatorlike lacewing larva makes short work of aphids. Both green and brown lacewings occur naturally in gardens; green species can also be purchased from garden suppliers.

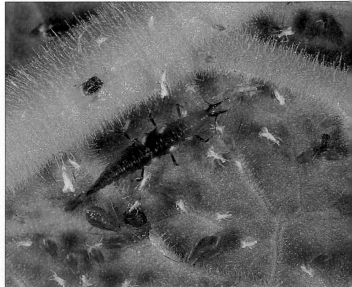

Soap sprays kill many kinds of soft-bodied insects and mites on contact by penetrating their cell membranes.

Armed with an arsenal of chemicals, gardeners were encouraged to spray or dust at the slightest provocation. The mere sight of an insect was just cause for an all-out war; no value was placed on preliminary observation and identification. Gardeners might not determine whether the creature they destroyed was a pest or a beneficial organism that attacked pests. They might never accurately identify their target—or even ascertain if the damage they aimed to correct was due to poor growing conditions or some other culprit.

Gradually, devotees of chemical control realized they were traveling down a treacherous road. Evidence mounted that various chemicals had adverse environmental effects. Overuse of pesticides was threatening the balance of nature. Broad-spectrum chemicals (those capable of killing many kinds of organisms) wiped out populations of beneficials while fostering strains of "super" pests resistant to pesticides. Some creatures that had not previously been pests became troublesome after pesticides eliminated the natural enemies that had kept them in check.

Faced with these problems, the experts turned from the notion of *eradication* to the idea of *management*. The focus was on minimizing pest populations rather than annihilating them, and on preventing problems by choosing appropriate plants and giving them good growing conditions. This approach to pest control is known as *integrated pest management (IPM)*. Since its introduction in the 1970s, IPM has been gaining currency among commercial growers—and it's now becoming popular in home gardens.

When action against a pest is warranted, IPM recommends that physical and biological controls be deployed first. Physical controls allow you to suppress a pest by strictly mechanical means, such as traps and barriers; biological controls involve releasing natural enemies and disease-causing microbes that attack pests. Chemical controls are used only as a last resort.

It should be noted that IPM isn't a purely "organic" approach; it doesn't outlaw the use of synthetic chemicals simply because they aren't "natural." There is minimal reliance on *all* chemicals, whether derived from nature or produced in the laboratory.

Fundamental to the whole concept of IPM is the notion that a certain amount of damage can and should be tolerated. There's no such thing as a pest-free garden, and trying to create one is unrealistic. You won't be able to do it, and you'll find the effort a frustrating waste of time.

Observation is at the heart of IPM. To know whether and when to act, you must inspect the garden regularly and monitor any problems. Using IPM principles is more time-consuming than just applying spray on a set schedule, but it's more effective in the end—and it encourages you to become attuned to your garden.

Chances are you've already begun your own IPM program without realizing it. If you've ever planted resistant varieties, blasted aphids with water, handpicked hornworms, trapped snails, or pruned a plant to improve air circulation, you've already instituted some elements of an IPM regime.

Commercial growers and pest-control professionals follow a structured program, but there are no rigid rules for home gardeners. This book will help you understand the basic principles of IPM and allow you to organize a regimen that works for you. You'll learn not only how to avoid problems, but also how to recognize trouble and what to do about it. You'll become familiar with a wide range of controls, many of them nontoxic. To help you identify troublemakers, the rogues' gallery on pages 71–110 offers photographs and descriptions of more than 100 different pests and diseases.

The effects of IPM may not become evident immediately—but rest assured, your garden *will* be healthier and more productive in the long run.

This gardener doesn't wait for trouble to find him, but instead heads off problems by inspecting plants regularly.

ℛECOGNIZING 𝒯ROUBLE

When you consider everything that can go wrong in a garden (for a hint, thumb through the rogues' gallery on pages 71–110), you may develop the sinking feeling that your corner of Eden is doomed to disaster. But there's no need to despair—though many garden problems are possible, few are likely to occur. Your plants won't succumb to every affliction, just as you won't suffer every calamity that can occur in an average day.

As you survey your domain for signs of trouble, be discerning: don't let the sight of a solitary bug or a few yellow leaves send you scurrying for the heavy artillery. A healthy garden is alive with insects, fungi, and other organisms. They're everywhere—in the air, on the soil, underground, and on plants. The great majority are harmless: of the few million species of insects and related creatures and the approximately 100,000 plant diseases, only a small percentage cause serious problems. Some creatures even help gardeners by pollinating plants, preying on pests, and breaking down organic matter. In fact, pests aren't responsible for most garden woes: your troubles can usually be traced to poor growing conditions (discussed further on pages 12–14 and in the next chapter).

Of course, you will have to do battle with pests and diseases at least occasionally,

so it's important to be aware of what's living in your garden—and to recognize the damage these organisms can cause.

Your region, climate, plantings, soil type, and gardening practices influence the types of trouble you'll encounter. Some problems are minor and can be monitored or entirely ignored, while others are serious enough to require immediate action. The challenge is to decide which of these categories your problem belongs to, and to determine its cause. For example, what if a plant that ordinarily remains vivid green turns a sickly yellow? Something is wrong, but what? Many gardeners blame the most visible creature, but the actual culprit may be microscopic; or the plant may be getting too much water or not enough nitrogen or light. Unless you figure out the underlying cause, you probably won't manage to cure the patient.

This chapter will help you assess your garden problems. The following pages offer information on the chief sources of trouble, tips for examining plants, and suggestions for identifying the villain (or at least narrowing the list of suspects) from the evidence collected. You may have to seek expert help in diagnosing some problems, but at least you'll know when that's necessary. And you'll be able to provide your consultant with valuable clues.

POTENTIAL CULPRITS

ost organisms in a home garden are either neutral or beneficial, but some can do serious harm. These troublemakers range from obvious creatures, clearly visible from a distance, to microscopic pests that are usually identified by the damage they do. The major sources of garden problems fall into the four categories reviewed in the following pages: insects and their relatives, larger creatures, plant diseases, and poor growing conditions.

Insects & Their Relatives

If the destruction in your garden is the work of a living creature, chances are that the guilty party is either an insect or one of the pests that look and act much like insects: spider mites (arachnids), slugs and snails (mollusks), pillbugs and sowbugs (crustaceans), millipedes (diplopods), and nematodes (roundworms). These pests invade your property in various ways: they fly, walk, or crawl in, or they're blown in by the wind or brought in on infested plants or soil.

The overwhelming majority of creatures in this group are insects. (Although they're often referred to as "bugs," that term is properly reserved for a specific group of sucking insects.) Insects comprise all those organisms which, as adults, have six legs, an exoskeleton (a protective outer shell instead of an internal backbone), and three body sections: head, thorax, and abdomen. The head contains the sensory organs, including eyes, antennae, and mouthparts. The legs and wings are attached to the thorax (most adult insects have two pairs of wings; flies have a single pair). The abdomen, usually the longest section, brings up the rear; it contains the respiratory, digestive, and reproductive organs.

A few million insect species inhabit the earth, with beetles accounting for some 40 percent of the total. Of all those insects, a mere one percent are considered pests. Moreover, because many insects change quite radically as they grow, pests are often harmful only at a certain stage of life.

How Insects Grow

Insects develop through the process of metamorphosis, meaning that their form changes as they mature. It's important to know a troublemaker's different stages, so you can recognize the pest before any damage is done. A basic knowledge of the life cycle will also tell you when a particular pest is most susceptible to

Simple metamorphosis

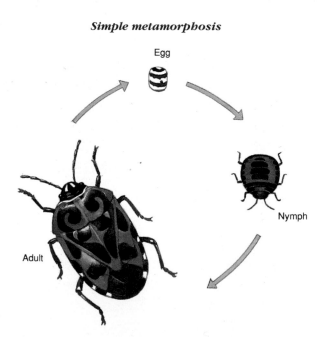

The harlequin bug is an example of a species that undergoes simple metamorphosis—meaning that the immature form looks pretty much like the adult. It starts out as an egg, then hatches into a wingless nymph, which increasingly resembles the adult with each molt.

Complete metamorphosis

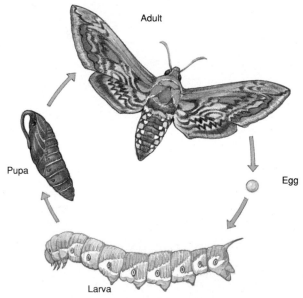

Species that undergo complete metamorphosis change drastically as they grow. The hornworm, illustrated above, starts life as an egg, then becomes a wingless larva called a caterpillar. After the last molt, it enters a resting stage called a pupa, then emerges as a moth.

controls. You'll find details on life-cycle stages—and on available controls for the different phases—in the insect profiles beginning on page 72.

Almost all insects begin life as an egg, though most aphids and a few unusual insects give birth to live young. The egg hatches into an immature insect called a larva, which grows by molting—periodically shedding its external skeleton. Because the exoskeleton is rigid, it can't expand to keep up with the insect's growth; when it becomes too tight, it splits open and the insect crawls out, having already developed a new, soft shell that will soon harden. Most insects molt a specific number of times during the larval stage.

In *simple (incomplete) metamorphosis,* the larvae (usually called nymphs) look pretty much like miniature versions of the adults. They molt until they reach maturity, growing larger with each molt. Development of wings, changes in color, and other modifications may occur during growth. For example, spider mites (which undergo simple metamorphosis just as some insects do) start out with six legs and develop two more. Nymphs and adults of the same pest species feed on the same plants and cause the same kind of damage.

In *complete metamorphosis,* the young insect is a wormlike creature that bears no resemblance to the adult. Moth and butterfly larvae are called caterpillars, beetle larvae are known as grubs, and fly larvae are called maggots. Caterpillars have legs, but maggots and some grubs do not. During the larval stage—the phase that's usually most damaging to gardens—an insect increases in size by molting. When it's ready to mature,

An insect increases in size by shedding its exterior skeleton. Having reached adulthood, this green cicada is molting for the last time.

METAMORPHOSIS & MOUTHPARTS

If you know what kind of mouthparts a pest has and what type of metamorphosis it undergoes, you'll have important clues for recognition. Pests experiencing simple metamorphosis look pretty much the same throughout life, while those undergoing complete metamorphosis change dramatically as they mature. The mouthparts determine the type of damage each pest inflicts. (For more on metamorphosis and mouthparts, see "How Insects Grow," facing page, and "Type of Mouthparts" on page 10.)

Pest	Metamorphosis	Mouthparts
Aphids, cicadas, leafhoppers, mealybugs, psyllids, scale insects, spider mites, spittlebugs, true bugs, whiteflies	simple	sucking
Crickets, earwigs, grasshoppers, pillbugs, sowbugs	simple	chewing
Slugs, snails	simple	rasping
Thrips	simple	rasping/sucking
Ants, beetles and weevils, caterpillars, sawflies	complete	chewing
Fly larvae	complete	rasping

it pupates; that is, it enters a transitional, nonfeeding phase during which it encases itself in a protective shell, such as a chrysalis or cocoon. A pupating insect is known as a pupa. Finally, after undergoing a dramatic transformation, the insect emerges as an adult.

Each stage of development may be lengthy or fleeting—some insects pass through the entire cycle in just days. Adulthood is usually very brief and concentrates on reproduction. Depending on the insect and the climate, there are from one to many generations a year; in mild-winter regions, some insects (such as aphids and whiteflies) reproduce the year around.

Once you understand a particular troublemaker's life cycle, you have the key to managing that pest. You'll find that most creatures have weak points, stages when you can more easily kill them. Entomologists refer to this as "breaking the life cycle." For instance, many insects that undergo complete metamorphosis are vulnerable during pupation, when they can't fly, crawl, or run away. At that point, you may be able to reduce the population by handpicking pupae attached to plants or garden structures, or by tilling the soil to destroy underground pupae or expose them to predators. If the damaging stage of a soil-dwelling pest feeds only on a certain host crop, you can starve the creature by rotating the crop (not growing it in the same place year after year).

Type of Mouthparts

Most insects and related pests harm plants by feeding on them. The mouthparts—usually chewing or sucking—dictate the kind of damage and provide clues for control. The larval and adult stages of insects that undergo complete metamorphosis often have different feeding patterns: the larvae inflict damage by eating plant tissue, while the adults just lap up or siphon pollen and nectar. In some cases, the adults live only long enough to reproduce and don't feed at all.

Chewing pests eat holes in leaves, twigs, stems, flowers, and fruit; some gobble right through and sever the stems. These pests have strong, sideways-moving jaws, and their teeth tear as well as mash. Caterpillars,

beetle grubs and adults, sawfly larvae, grasshoppers, and earwigs all chew. A few types of maggots are also chewing pests, but most fly larvae have a pair of hooks they use to rasp or scrape food. Among insect relatives, pillbugs and sowbugs have chewing mouthparts. Slugs and snails, though sometimes described as chewers, actually consume food with a radula—a tooth-encrusted band that shreds food and draws it into the mouth.

Some chewing pests leave distinctive holes or other marks. Root weevil adults make notches in leaf edges. Flea beetle adults munch small, irregular holes, so that foliage looks as if it's been riddled with shotgun fire. Many other beetles skeletonize leaves, chewing them to lace.

Insects that chew tunnels in stems and trunks are called borers. They cause plants to yellow, wilt, and suddenly lose vigor; branches or stems die back. And unlike most insect pests, borers don't just disfigure woody plants—they often kill them. Of the numerous types of tree borers, many have characteristic tunneling patterns, allowing experts to identify the cause of damage even if the culprits have already vacated. Other types of borers attack nonwoody plants, such as squash and corn. To recognize borer holes in plant stems, look for excrement at the hole edges and on the ground below; on woody plants, sawdust or sap may also be present.

Sucking pests cause leaves, buds, and fruit to discolor, distort, or drop, but they don't cut away pieces of the plant. Most pests of this type insert a

The irregular notches chewed in the edges of these rhododendron leaves were made by adult root weevils.

Curled leaves and shriveled fruit on this apple tree were caused by sucking pests, specifically aphids. Sucking removes sap from sections of plant tissue, thus distorting it.

Some chewing insects damage plants by tunneling inside the stems. This stem has been cut open to reveal a squash vine borer.

specialized feeding tube (called a stylet) directly into a plant's vascular tissues, then suck the juices—sometimes transmitting diseases as they feed. Sucking pests include aphids, scale insects, mealybugs, whiteflies, leafhoppers, psyllids, true bugs, spider mites, and thrips. Thrips are unique in rasping or scraping the leaf surface before sucking the juices.

Numerous means of control are possible for most pests (see "Dealing with Pests & Diseases," pages 41–69, for details), but mouthparts play a role in determining the chemical controls that will be effective. Since chewing and rasping pests ingest plant material, they can be killed with a stomach poison applied before or during feeding. Sucking pests, on the other hand, don't eat outer surfaces, so poisons that must be ingested along with plant tissue won't do the trick. More effective are contact poisons, which kill by asphyxiating or paralyzing the pest when it's directly hit. Or apply a systemic poison—one that's absorbed into a plant (use it only on ornamentals). Both systemic and contact poisons also work on chewing pests that feed on plant surfaces. Once they're inside a plant, borers are hard to control with poisons; other methods, such as injecting parasitic nematodes (see page 54) into the bore holes, are more efficient.

Larger Creatures

Urban sprawl has brought people into direct contact with wildlife. Consequently, many home gardeners now find that the most serious pests aren't insects, but larger creatures such as birds, rabbits, deer, and other fauna. Most of these pests are easy to recognize if you know when and where to look for them (see pages 93-97), and many cause highly distinctive damage. Some nibble fruits and vegetables just before harvest, others girdle trees and shrubs by gnawing, and yet others heave plants out of the ground by burrowing. And, even a pet dog or cat can wreak havoc by trampling plants and digging up seedlings.

Doing battle with wild creatures seems to cause gardeners unbounded frustration. When you feel your blood pressure rising, it may help to remember that you're the one infringing on the animals' territory—they're only doing what comes naturally. However, tolerance and understanding go only so far where some pests are concerned. A single mole or pocket gopher, for example, can move through a garden like a wrecking crew. Trapping to kill may be the best way to resolve this kind of problem, but deterrence is a better solution in many cases. Bird netting, row covers, fencing, tree guards, and repellents are all available weapons in the war against wildlife.

If you do decide that an animal is too troublesome to bear, consider its legal status before taking deadly action. Depending on the area, larger mammals (such as deer, rabbits, squirrels, and raccoons) may be protected except during official open seasons. For information, check with your state game or conservation department.

Plant Diseases

Plant diseases are caused by pathogens—primarily fungi, bacteria, and viruses. Unable to manufacture their own food as green plants do, these organisms obtain sustenance from host plants, causing disease in the process. Luckily, most plants are resistant to a wide range of disease organisms, but just about every plant is vulnerable to some ailments.

The presence of a pathogen in your garden doesn't mean that disease is sure to develop. Problems will arise only if the plant is susceptible to the pathogen *and* if the environment favors development of the disease. For example, if a plant is susceptible to a particular fungus, but soil conditions are unfavorable, the disease won't appear. If the soil conditions are conducive but the plant is resistant, the disease still won't develop. This interrelationship is known as the disease triangle.

As experienced gardeners know, avoiding plant diseases is far easier than trying to cure them. To dodge a disease, you disrupt the disease triangle (this is equivalent to breaking an insect's life cycle by exploiting its vulnerabilities). Since eliminating the pathogen isn't always easy, your best bet is to use resistant plants (if they're available) and to provide growing conditions that discourage the disease. If the pathogen is favored

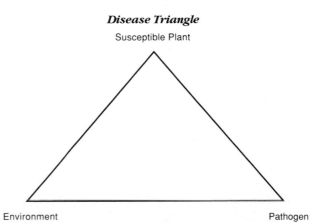

Disease Triangle
Susceptible Plant

Environment   Pathogen

Three elements must be present for a plant disease to develop: a disease-causing agent called a pathogen; a plant susceptible to that pathogen; and an environment (such as weather or soil conditions) favorable to development of the disease.

Fungal structures, such as this orange growth, are often quite conspicuous.

by wet soil, let the ground dry out between waterings; if it needs shade, prune the plant to let in sunshine.

Temperature and humidity are key environmental factors, but you can't do much to control them. As long as the weather is unfavorable, the disease will be slowed or even stopped in its tracks. But the longer the conditions remain ideal for disease development, the more severely your susceptible plants will be affected.

Fungi

The source of about 80 percent of plant diseases, these threadlike organisms develop structures that actively penetrate plant tissue—and that are often plainly visible on infected plants. Fungi usually grow through or on the tissue as fine branching filaments (hyphae), which form a mass of strands (mycelium). Fungal growth is usually favored by warm, humid, or moist conditions. Infected plants typically suffer from rotting, stunting, leaf curling, spotting, and wilting.

Many fungi produce tiny reproductive bodies called spores, which spread around the garden by means of wind, water, insects, infected soil and tools, and gardeners moving from one plant to another as they work. If a spore lands on a suitable host, it germinates when conditions are right, producing a new infection. Some types of reproductive structures can survive for long periods in the soil, even in the absence of a host organism.

Bacteria

Because these single-celled microorganisms generally need high temperatures and moisture to multiply, the diseases they cause are more common in the tropics than elsewhere. Bacterial diseases are relatively uncommon in dry-summer climates (fireblight is a notable exception), although overhead irrigation can create favorable conditions. Common symptoms of bacterial disease include wilting, rotting, and swollen plant tissue (galls). Bacteria often live in a protective ooze they produce in the infected plant.

Unlike fungi, bacteria don't have an active mechanism for penetrating tissues: they enter their victims by slipping through natural openings and wounds. Splashing water is the most common method of transmission, although insects, infected tools and soil, and gardeners working among plants can also foster the spread of disease.

Viruses

Even smaller than bacteria, viruses invade living plant tissue and reproduce inside the cells. They're most commonly spread by sucking insects (especially aphids, leafhoppers, whiteflies, and thrips), but can also be transmitted by infected seeds, tools, and hand contact.

Viruses cause stunting, malformations, and color changes, most often mottling or yellowing. Some attractive plant varieties—striped tulips, for example—owe their variegation to a viral infection. Viruses can be disastrous for commercial growers, but they aren't usually a problem for home gardeners, especially when plant vigor isn't affected.

Since viral growth is linked to that of the host plant, you can't get rid of viruses without damaging or killing the host as well. In theory, you can stop many viral diseases by eliminating every potential carrier insect, but this method of control is impractical. You can, however, keep the disease from spreading by destroying infected plants.

Poor Growing Conditions

Though the blame is frequently directed elsewhere, most garden trouble results from stress due to unfavorable growing conditions—including nutrient deficiencies (see facing page), poor drainage, improper light, temperature extremes, air and water pollutants, fluctuations in soil moisture, chemical damage, and mechanical injury. In fact, adverse conditions are often responsible for common problems such as yellowed leaves and stunted growth, ailments many gardeners automatically attribute to insect pests or plant diseases. These problems are sometimes referred to as environmental diseases to differentiate them from sicknesses in which a pathogen is involved. They're also called physiological disorders.

Stressful growing conditions can be categorized as acute or chronic. Acute stress, brought on by sudden, short-term problems such as an untimely freeze or improper pesticide application, produces immediate damage. Chronic stress, caused by long-term problems

NUTRIENT DEFICIENCIES

When a plant doesn't get the nutrients it needs from the soil, it shows symptoms of the shortage. Since some of these symptoms can also result from pest or disease attack, inspect the plant and its growing conditions carefully before concluding that the problem is nutritional.

The symptom most frequently noticed by gardeners everywhere is leaf yellowing due to nitrogen deficiency. Because nitrogen is needed in large quantities and leaches from the soil easily, it must be replenished on a regular basis. The other nutrients are either required in small amounts or don't easily move through the soil, so they're added only periodically or when a deficiency is noted. (For more information about building nutrient-rich soil, see "Building Good Soil," pages 22–25.)

Element	Symbol	Symptoms of Deficiency
MAJOR NUTRIENTS		
Nitrogen	N	Slow growth and stunting; uniform yellowing, starting on older leaves and spreading to whole plant; dropping of oldest leaves.
Phosphorus	P	Slow growth and stunting; poor flowering and fruiting; purpling of leaves on some plants.
Potassium	K	Slow growth; weak stalks; browning on tips and edges of older leaves; small fruit or shriveled seeds.
SECONDARY NUTRIENTS		
Calcium	Ca	Poor growth; death of growing points.
Magnesium	Mg	Yellowing between the veins on older leaves, with dead spots appearing suddenly; curling of leaf edges.
Sulfur	S	General yellowing of leaves; stiff appearance of plant.
MINOR NUTRIENTS		
Zinc	Zn	Yellowing between the veins on young leaves; abnormally small new leaves.
Iron	Fe	Yellowing between the veins on young leaves, with small veins also turning yellow.
Manganese	Mn	Yellowing between the veins on young leaves, with all veins remaining green; black spots next to veins.
Copper	Cu	Stunted growth; sunken dead spots starting on youngest leaves; wilting of leaf tips.
Boron	B	Death of growing tips, causing lateral buds to develop and produce fans of stiff shoots ("witches' brooms"); brittle stems and other plant parts.
Molybdenum	Mo	Pale yellowing between the veins, starting on older leaves and spreading to whole plant; upward curling of leaf edges.
Chlorine	Cl	Wilting, followed by loss of color.

Air pollution is responsible for the speckled leaves on these potato plants.

such as a nutritional imbalance or insufficient light, causes more gradual damage. In either case, though, the affected plants grow weaker and may even die.

Stress also opens a plant to potentially fatal invasions by pests and diseases. Bark beetles and borers will often strike a tree weakened by poor growing conditions; one that's been gashed by a lawn mower or string trimmer is more susceptible to disease, since the wound offers easy access to pathogens. Crowded or overwatered seedlings are more likely to be killed by damping-off fungi than those raised under more favorable conditions. Pests and diseases may administer the coup de grace in such cases, but they aren't the real source of trouble.

Poor growing conditions often produce symptoms similar to those typical of diseases, but the treatment that's required differs. A hint for telling the two types of problems apart: if the same symptom (such as wilting or yellowing) appears on several adjacent plants, chances are the cause is environmental. Disease damage is usually distributed randomly or limited to a single plant. Unless a disease is obvious, it's usually easier to assume that growing conditions are at fault—and to see if improving them solves the problem.

Here's a short list of environmental problems often blamed on pests or diseases:

■ Cracking or splitting of tomatoes, cherries, and other fruits may occur when the plant receives fluctuating moisture.

■ Blossom-end rot on tomatoes (a dark, sunken area on the bottom of the fruit) and brown spots or a brown core inside apples are among the problems due to calcium deficiency. If the soil is too wet or too high in salts, or if the moisture content fluctuates, calcium uptake may be prevented even when the nutrient is present in the soil.

■ Catfacing (distortions on the bottom of tomato fruit) results if weather is too cool when the plant is young.

■ Misshapen cucumbers and apples can often be attributed to poor pollination due to unfavorable weather or the absence of bees.

■ Blasting (premature dropping of buds and flowers) is caused by soil that's overly wet or dry.

■ Bleached or dead tissue between leaf veins is a sign of sunburn—a problem that usually afflicts plants exposed to hot sun and given insufficient water.

■ Dead tissue at leaf margins may be due to windburn, a common problem when plants growing in dry soil are subjected to dehydrating winds. The same symptom can result if plants receive excess salts from overfertilizing or from lime or other minerals in the soil or irrigation water.

Uneven watering is the usual cause of splitting in carrots. Forking may be due to hard or rocky soil; it can also result when roots come in contact with fresh manure.

■ If trees and shrubs in a big city drop their leaves prematurely and show yellow or purple areas between the veins on older leaves, the culprit could be too much ozone in the air.

TRACKING DOWN TROUBLE

ou have to see trouble in order to unmask it. Unless you conduct regular patrols of your garden, small problems can mushroom into big ones while you're not looking. Walk among your plants often, using a 10-power hand lens for a close look at small organisms and plant parts. Keep a written record of all the problems you discover, noting the date, symptoms, and any unusual conditions, such as unseasonal weather or a recent pesticide application.

When you find what you think is trouble, don't jump to conclusions. Before making a diagnosis, ask yourself what the plant should look like. It may turn out that yellow leaves are normal, or that the shrub you think is stunted isn't supposed to grow more than a foot high. Leaf drop can be natural, too: deciduous plants drop all their leaves at once, while evergreens lose some of their older leaves throughout the year. As is true for deciduous leaves, evergreen leaves sometimes turn brilliant colors before they fall.

Fruit trees normally thin themselves by dropping fruit. Some varieties don't flower or bear fruit at all until several years after planting; others bear a heavy crop one year and little or no fruit the next, while still others produce fruit only when another pollinating variety is present. Certain plant species, such as holly, have separate male and female plants, so find out whether your plant is a male before you fret over a lack of fruit.

Once you decide that something is amiss, make sure you've properly identified the plant. This may help you track down the culprit, since some pests and diseases attack only a single species or a narrow range of plants. Closely related plants are usually attacked by the same organisms—for example, cole crops are susceptible to imported cabbageworms, while members of the rose family are prone to the bacterial disease fireblight. Identifying a plant correctly can also help you eliminate suspects; it may be that some pests eat just about everything *but* the plant in question.

Many pests and diseases will appear in your garden at the same time each year. Keep track of their comings and goings; you may be surprised by the regularity of their visits. Certain insects can be counted on to show up within a few days of a specific date each year. Some disease organisms make inroads at particular times—during a wet spring, for example. By learning to recognize these habitual offenders, you'll be able to anticipate their arrival and have your defenses ready.

Friend or Foe?

It's important to perform general surveys of your garden at regular intervals. Find out what's sharing your yard with you—and remember that not all insects and other organisms you encounter are harmful. Learn to recognize beneficial creatures (see pages 36–39) and treat them with respect. These helpful organisms are efficient pest-control agents; kill them, and your job becomes more difficult.

Some people view anything that creeps, crawls, or slithers as a pest, but many of these creatures are garden allies. Lizards and salamanders eat insects; garter snakes and turtles devour slugs. Ladybird beetle larvae and lacewing larvae, both of which resemble tiny alligators, feed voraciously on aphids and other pests, as do the wormlike larvae of syrphid flies.

The larval form of the mealybug destroyer (top) bears a striking resemblance to its prey, the mealybug (bottom).

Other types of creatures unjustly maligned include birds, bats—and sometimes, pets. But bats consume huge numbers of insects; and though some birds are a nuisance, many species dine almost entirely on insects, ignoring seedlings and ripe fruit. Even a well-trained cat can provide invaluable aid to gardeners by hunting rodents such as pocket gophers and voles.

Some creatures are potentially harmful, but rarely appear in large enough numbers to cause trouble. The ferocious-looking Jerusalem cricket, which eats potato tubers, doesn't usually pose a serious problem in home gardens. Two insects that are hard to spot because they look like plant parts—katydids (leaf mimics) and walkingsticks (twig mimics)—feed on foliage but usually don't do much damage. Don't wage war on these "pests" until they give you reason to do so.

In other cases, an ally and a pest may look alike, so make sure you've got the identification right before you put out the welcome mat or launch controls. You'll find that ladybird beetles bear a superficial resemblance to several harmful insects: Mexican bean beetles, spotted cucumber beetles, and Colorado potato beetles. Except for its thicker middle, the moth of the destructive peachtree borer looks like the digger wasp, a helpful insect that stocks its burrow with caterpillars and other pests.

Some creatures are capable of doing harm but rarely inflict it. Among them are two masters of camouflage: the katydid (left), which looks like a leaf, and the walkingstick (right), which resembles a twig.

The wax-coated larvae of the mealybug destroyer bear a remarkable resemblance to their quarry—mealybugs. If it weren't for a wider body and oversize orbs, big-eyed bugs could be mistaken for the chinch bugs they prey on. The similarity between predatory mites and spider mites ends with their looks: the former are garden allies, the latter enemies. And while it's true that most thrips harm plants, the six-spotted thrips and a few other species are predators.

The different dining preferences of brown garden snails and decollate snails prove that not all snails are bad. Brown garden types ravage plants; decollates eat their brown cousins. If you have both kinds of snails in your garden, make sure you don't kill the wrong ones.

If you dig up a plant and find nodules on the roots, don't panic if the plant is a legume (such as a bean or pea) and you can flick off the growths—those little lumps are nitrogen-fixing nodules, which help the plant convert nitrogen into a usable form. On the other hand, if the plant isn't a legume or if you can't detach the growths, you're probably looking at root galls caused by pest nematodes.

Not all garden creatures are undesirables, and the same holds true for plant diseases. Some pathogens actually help gardeners. Naturally occurring viruses commonly kill pests such as cabbage loopers, imported cabbageworms, codling moths, and armyworms. A viral or fungal disease reduces gypsy moth populations every 10 years or so. Chinch bugs are another troublemaker susceptible to fungal infection. If you see sick-looking or dead pests, leave them alone to infect other members of the species. Or collect the diseased individuals and turn them into a pest spray; see page 57.

Examining Plants

Use the following technique to inspect a plant, making notes of your findings as you go. From the accumulated data, you may be able to zero in on the culprit yourself. If you're baffled and decide to consult a professional, at least you'll have useful information to convey.

Start your examination at the bottom of the plant, since damaged roots cause many aboveground symptoms. Check the soil at the root zone to find out if it contains adequate moisture: either dig down with a trowel or insert a soil probe. Don't just check the surface, since that dries out first.

Uncover a section of root and follow it to the end to see if it's rotten or infested. Dark and/or smelly roots often indicate soggy soil or root-rotting organisms (healthy roots are usually whitish and don't have a foul odor). Look carefully for any pests. Notice whether the roots are chewed off or damaged in some other way.

Next, check the crown (where the roots meet the stem) for signs of pests or diseases. Peel away any bits of loose or wet bark to see what's underneath. Note any discoloration or unusual odor.

Work your way up the stem and branches, searching for wounds, nicks, holes, and dieback. Wounds at the bases of trees are often caused by lawn mowers or gnawing animals. Holes with sawdustlike material at the edges indicate the presence of borers. Check the junctions of branches and stems for anything unusual.

Examine the leaves—especially the undersides, where many pests feed. Note any twisting or curling, stickiness, or abnormalities in leaf color, size, or vigor. Look for spots and holes, too.

If flowers or fruit are present, inspect them for signs of infestation or other damage, such as spotting or malformations. If the plant is supposed to bloom or bear fruit but hasn't, note that.

Check to see whether the symptoms appear over the whole plant or are confined to just one section. Also try to discern whether some of the damage appears fresher than the rest. Generally, the most recent damage is characteristic of the underlying problem; older damage may in part be attributable to secondary pests or disease organisms.

Identifying the Culprit

Since different problems have different remedies, an accurate identification of the troublemaker is essential. The evidence you've gathered from touring the garden and examining ailing plants should help you uncover the scoundrel—or at least close in on it.

Before you attribute blame to an organism, make a final check of growing conditions. Might they be the source of trouble? Even if harmful organisms are present, they're not necessarily the culprits—they may simply have taken over a stressed plant rather than instigating the problem. If you can improve conditions, do so. If the environment looks good, then focus your attention on pests and diseases.

Fortunately, only a limited number of troublemakers are likely to show up in any garden, so there's no need to go through a mental checklist of every conceivable pest and disease. Consider only those found locally and known to affect the plant in question.

If you've found a pest on the plant, then your task is to put a name to it. Keep in mind, however, that the presence of a pest that *could* have inflicted the harm doesn't constitute proof that it *did* do the damage. For example, sowbugs and pillbugs, largely beneficial creatures, often hide in holes chewed by slugs and snails. The most conclusive evidence, of course, is seeing the pest actually feed on the plant. To catch nocturnal pests in the act, you'll have to equip yourself with a flashlight for nighttime surveillance.

If no troublemaker is in evidence (during day or night), the best course is to run through the list of suspects to select the most likely perpetrator. You may already be familiar with some pests and diseases, and you'll learn about others by talking to knowledgeable gardeners in your area. Pages 72–105 of this book offer profiles of over 100 pests and diseases; to narrow the search more rapidly, you can check for the afflicted plant in "Some Common Plantings & Their Pests" (pages 106–110), then start by investigating the troublemakers listed for that plant. You'll notice that some organisms have distinctive features that make them easy to recognize. Others are usually identified by the damage they do or the telltale signs they leave.

Among the most easily recognized pests are Colorado potato beetles, which sport a polka-dot thorax and striped wing covers. You'll know earwigs by the curved pincers at their back ends. Root maggots can be identified by their pointy heads and blunt rears; a magnified look at an aphid reveals two projections at the rear that resemble dual exhaust pipes. If tiny, white mothlike creatures flutter up in a cloud when you touch a plant, you're looking at whiteflies.

The eggs of some pests are just as unmistakable as the creatures themselves. Harlequin bug eggs, for example, look like rows of tiny black and white barrels on leaf surfaces. Slugs and snails lay masses of gelatinous eggs in the soil. Fall cankerworm moths deposit gray, flowerpot-shaped eggs on twigs; imported cabbageworm butterflies lay bullet-shaped, ridged yellow eggs on the undersides of cole crop leaves.

Some symptoms obviously point to certain diseases. Blackened foliage and twigs on a pear tree scream fireblight; puckered leaves on a peach tree indicate peach leaf curl. Light pink slime oozing from lesions on plant parts means anthracnose. Tumorlike growths on the roots or stem near the soil line signify crown gall.

Even if you don't recognize a pest or disease, you may be able to narrow down your choices. For example, a powdery or fuzzy coating on plant parts is a clear sign of fungal disease. Weevils (a type of beetle) have a distinctive long, curved snout. All true bugs, many of which are pests, have a triangular marking on the thorax in the adult stage. An organism's color, shape, size, and favored host plants can also help you home in on the culprit. Weather conditions and time of year may prove significant as well.

You don't need to see some pests to recognize their handiwork. Pocket gophers and moles make distinctive mounds—fan-shaped mounds in the case of gophers, volcano-shaped ones for moles. A mole leaves the hole at the mound's center open, while a gopher plugs its off-center hole (although the opening may be unplugged if the gopher's work was interrupted).

Deep, irregularly spaced holes that slope down at an angle indicate a tree borer, while shallow holes in neat rows are the work of a less damaging creature—the sapsucker (a bird). Leaf miners create winding trails or blotches on leaf surfaces. Bagworms weave sacks that dangle from tree branches.

Other calling cards include the silvery slime trails on which slugs and snails travel. Fall webworms and tent caterpillars produce webs; webworms work at the ends of branches, tent caterpillars in the forks. Spittlebugs manufacture froth. Plum curculios leave a crescent- or mushroom-shaped scar on fruit. Hornworms are often hard to spot on a plant, but the large pellets of excrement they leave on the ground or foliage beneath skeletonized leaves give them away.

If you see leaves coated with a sticky material or black fungus, or if you notice ants marching along a plant's stem, you know you're dealing with sap-feeding creatures such as aphids, scale insects, mealybugs, whiteflies, leafhoppers, or psyllids. These pests excrete honeydew (undigested plant sap), which in turn attracts ants and promotes the growth of sooty mold.

What If You're Stumped?

Don't expect to figure out every problem yourself. A single symptom may have many possible causes (see the chart on pages 18 and 19); some organisms are

MAJOR SYMPTOMS & POSSIBLE CAUSES

Sometimes the cause of a garden problem is easy to identify. Fan-shaped mounds of soil littering a lawn are obviously the work of a pocket gopher; slime trails leading to plants with large, ragged holes in the leaves can only mean slugs or snails.

A bigger challenge is deducing the source of trouble when many explanations are possible. Leaf yellowing and wilting, for example, may be due to insects, nematodes, diseases, or improper watering.

If you can't find clear evidence of a pest, the best course is to assume that growing conditions are at fault, then to focus on improving them. If you're convinced a pest or disease is the villain, check the rogues' gallery (see pages 71–110) for a likely candidate—but be sure *all* the evidence fits before placing blame. For help in solving problems you can't identify on your own, consult your Cooperative Extension office or a reputable local nursery.

Symptoms	Possible Causes	Notes
Leaf yellowing	Spider mites	Usually pinpoint yellowing
	Insects	Especially bark borers
	Nematodes	Wilting also occurs
	Root rot diseases	Older leaves affected first
	Viral diseases	
	Overwatering	Older leaves affected first
	Nutrient deficiencies	See page 13
	Soil chemistry	
	Old age	
Leaf mottling	Viral diseases	
	Pesticide damage	
	Natural on some plants	Genetic variegation
Leaf spots	Diseases	
	Insects	
	Spider mites	
	Air pollution	
	Pesticide damage	
	Nutrient deficiencies	See page 13
	Soil chemistry	
Leaf malformations	Sucking pests	
	Viral diseases	
	Peach leaf curl	On peaches and nectarines
	Lack of water	
	Nutrient deficiencies	See page 13
	Pesticide damage	Especially from certain weed killers
Holes in leaves	Chewing pests	
	Birds	

difficult to detect, while the presence of others can't be confirmed without laboratory analysis. The rogues' gallery (pages 71–110) contains a good cross section of pests and diseases, but it certainly doesn't cover all the possibilities. Even if you can't name the culprit, though, you'll probably be able to whittle down the list of candidates—especially if you've correctly identified the damaged plant.

There's nothing wrong with making an educated guess: that's the way to learn. But confine your guessing to plants you can afford to lose. Guess wrong about an ailing tomato or zinnia and you haven't lost much, but make a faulty assumption about a mature oak or elm tree and you risk losing a great deal.

Your Cooperative Extension office, the County Agricultural Commissioner's office, reputable local

Symptoms	Possible Causes	Notes
Leaf drop	Diseases	
	Bark borers	
	Improper watering	
	Improper light	
	Air pollution	
	Nitrogen deficiency	
	Soil chemistry	
	Natural on some plants	Deciduous plants
Cankers (sunken lesions)	Diseases	
	Insects	Especially borers
	Root stress	
	Sunburn	
	Mechanical wounds	From lawn mowers, pruners, and other equipment
Branch dieback	Diseases	Especially vascular diseases such as fireblight and verticillium wilt
	Insects	Severe infestation by root or foliar pests or twig girdlers
	Nematodes	
	Root stress	
	Soil chemistry	
Root rot	Diseases	Usually fungal
	Overwatering	
	Root stress	
	Soil chemistry	
Discolored tissue inside roots and stem	Wilt diseases	
	Fertilizer damage	
Root malformations	Nematodes	
	Diseases	Crown gall, for example
	Girdled roots	Formed in container
	Natural on some plants	Nitrogen-fixing nodules on legumes
Poor germination	Old seeds	
	Uneven irrigation	
	Incorrect planting depth	
	Damping-off disease	
	Soil crusting	
	Insects	Root maggots, wireworms
	Birds	
Wilting	Lack of water	
	Insects	Especially borers and root feeders
	Nematodes	
	Diseases	Root rots and wilt diseases

nurseries, and experienced garden-club members are usually good sources of help. An accredited tree care specialist (an arborist) can assist you in diagnosing and treating trees. Some universities, botanical gardens, and garden centers sponsor "pest days" when home gardeners can bring in problems for diagnosis.

Rather than simply describing a plant's symptoms, bring actual specimens in a plastic bag or glass jar. If you can, pull up the whole plant. Otherwise, cut off a sample of the damaged part (include healthy adjacent growth). If you find an insect or other organism on or near the damaged plant, enclose that too. The best clues are often found at the junction of sound and damaged tissue. The samples should be fresh, since decaying or dead plant material is often overrun by secondary organisms, making diagnosis difficult.

Foster a healthy garden by eliminating weeds, which compete with desirable plants for water, nutrients, light, and space.

PREVENTING PROBLEMS

The easiest way to suppress trouble is to create a healthy garden. When plants get what they need for sound growth, they have the ammunition to fight pests and diseases on their own, largely freeing you from pest-control battles.

To find out what the plants in your garden require, consult a good gardening encyclopedia or a reliable local nursery. You don't have to fulfill every need with scientific accuracy; if you provide the necessary nutrients and growing conditions within a certain range, most plants will perform well.

Prevention of problems begins with groundwork. A deep, rich, highly organic, mulch-covered soil gives rise to sturdy, robust plants that hold their own against evildoers.

Choosing the right plants is important, too. Select types adapted to your growing conditions—and if they're resistant to certain pests or diseases prevalent in your area, so much the better. Depending on their degree of resistance, such plants can usually be counted on to evade trouble—or, at the very least, to survive it without suffering too much damage.

Trickery is another tactic in the gardener's arsenal. Outmaneuver pests by sceding or setting out plants earlier or later than usual, and by rotating crops from one place to another each year to deprive soil-dwelling organisms of the food they need to survive. Of course, proper planting techniques are essential, or all your scheming and plotting will be wasted.

Laying the foundations for good garden health is crucial—but you need to take steps to ensure that your plantings stay healthy, too. This chapter offers information on important maintenance activities such as watering, fertilizing, and cleaning up. You'll also learn how to enlist beneficial creatures as garden allies; study the photographs and descriptions of common beneficials on pages 36–39, so you'll recognize helpers when you see them.

Be aware that preventing problems takes time and work, but the effort pays off in healthier plants and a greater insight into the workings of your garden. If you devote extra hours to plant inspection, soil amendment, and other preventive practices, you'll ultimately spend fewer hours fighting pests and more time just enjoying the garden.

STARTING FROM THE GROUND UP

ake preventive practices an integral part of gardening, beginning with good soil preparation: careful groundwork can mean the difference between weak, ailing plants and strong, healthy ones. If you wait until the garden is up and growing to start thinking about prevention, you'll be too late to head off problems.

Building Good Soil

It's no exaggeration to say that good soil is the very basis of a thriving garden. A deep, rich, crumbly soil with a high organic content is able to hold a consistent supply of water, air, and nutrients, allowing plants to grow steadily—not in spurts, as they do if raised in poor, unproductive soil and forced to rely on periodic feedings and waterings for sustenance. Growing in fits and starts weakens most plants and increases their susceptibility to pests and diseases.

Any soil is improved by regular additions of organic matter. If you use wood products, make sure they're well composted or fortified with nitrogen; if they aren't, add nitrogen when digging in the amendment, then wait about a month before planting.

All good soils have plenty of spaces between individual particles. Besides serving as reservoirs for air and water, these spaces provide channels for root growth. Earthworms are attracted to crumbly organic soils, and their networks of tunnels create additional passages for roots.

Each pound of good soil is home to billions of microorganisms, many types of which are important for a garden's well-being. As the microbes break down organic matter, they convert nitrogen and other nutrients into a form that roots can absorb. Certain microbes have an even more direct role in plant health. Some manufacture antibiotics and other protective substances; others suppress the soil-dwelling fungi responsible for such diseases as damping-off and water mold root rot (see pages 99 and 105).

Soil Types

There are many types of soils, but gardeners usually talk about just three: clay, sand, and loam.

Clay soils hold water and nutrients well, but their fine, flattened particles pack together so tightly that root growth and drainage are hindered. Take care in irrigating plants in a clay soil: if you apply water too fast, it will run right off, but if you soak the soil thoroughly, plant roots may suffocate before the water drains away.

Sandy soils are made up of relatively large, rounded particles that fit together loosely—so loosely that water gushes right through. For this reason, plants growing in sand need frequent watering (and fertilizing, since sand retains nutrients poorly). On the positive side, sandy soils hold plenty of air for good root growth.

The ideal soil is *loam,* which combines the favorable traits of sandy and clay soils: retention of moisture and nutrients, good aeration, and fast drainage.

Amending the Soil

Short of replacing all your soil with loam, the easiest way to obtain good soil is to add organic matter regularly. Humus—the substance that remains after organic material decomposes—restructures soil by pulling the mineral particles together into spongy aggregates. It helps sandy soils retain moisture and nutrients; it opens up clay soils, improving air flow, water penetration, and drainage. And if you already have a good, loamy soil, organic matter helps maintain it.

Note: If you suspect your soil has problems that can't be solved by adding organic matter, consider sending off a sample for laboratory analysis. In some states, soil testing is offered through the Cooperative Extension Service; in others, you must deal with a soil laboratory (look for one in the Yellow Pages).

SOLARIZING THE SOIL

At the height of summer, take advantage of the sun's energy to clean up your soil through solarization. You simply spread a clear plastic sheet over cultivated soil, thus trapping the sun's radiant heat and raising the soil temperature high enough to destroy many harmful organisms and weed seeds.

Among the disease organisms easily killed are the fungi causing crown gall, damping-off, fusarium wilt, southern blight, and verticillium wilt (see "Plant Diseases," pages 98–105). Nematodes (see page 85) are tougher targets, but can usually be controlled well enough to improve the growth of shallow-rooted annual flowers and vegetables. Most weeds and weed seeds will die, although a few heat-resistant types, such as Bermuda grass and red clover, usually survive. Earthworms won't be harmed; they can escape danger by tunneling deeper.

Soil solarization works in most areas where the weather is relatively clear for a 4- to 8-week stretch during the hottest time of year. The higher the temperatures and the longer the plastic is left on, the better the results will be; in hot climates, the benefits can last for up to 3 years. Solarization isn't as effective in coastal climates, since fog, cloud cover, and wind interfere with the transfer of heat.

Because the soil surface must be exposed to the sun, don't try solarizing soil under a shade tree or against a tall fence that blocks sunlight. The bed to be solarized should be at least 2 feet wide; if it's any narrower, it will lose too much heat at the edges.

Cultivate the bed and prepare it for planting. If you plan to install an underground irrigation system, do so before solarization to avoid disturbing the soil later. Smooth and level the soil, then wet it a foot deep to improve heat transfer.

The traditional method calls for covering the bed with a single sheet of clear (not black) 1- to 4-mil plastic. However, two sheets separated by about 2 inches work even better. Spread one sheet on the ground; then drape the other over spacers, such as soda cans (lying on their sides) or pieces of PVC pipe. Seal the edges of the plastic with soil to keep heat and moisture in. Sweep off any water that collects during rainstorms. If rips develop, patch them with clear duct tape.

Once you remove the plastic, try not to contaminate the clean soil. Avoid moving soil or plants from untreated areas into the solarized bed; also try to plant within the top few inches of soil.

Edges buried in soil

1- to 4-mil plastic

Soda can spacers between 2 plastic sheets

Types of amendments. Various organic materials are suitable as soil amendments. Many, such as dry leaves and grass clippings, are readily available from your own garden. Homemade compost (see facing page) is a classic amendment, and it's still one of the best. Other materials, such as ground bark and animal manures, are sold at garden centers and soil yards. Crop by-products, including ground corncobs, grape pomace, peanut hulls, and cocoa bean hulls, are marketed in areas where those crops are grown.

If you plan to amend your soil with uncomposted manure or wood products, you'll need to take a few precautions. Fresh manure can burn plants, so add it to the soil several months before planting (or let it stand for a few months before you dig it in). The organisms that break down wood products require nitrogen to fuel their activity and will take what they need from the soil, depriving plants of this essential nutrient. Packaged and some bulk wood products are usually nitrogen-fortified, so they can be applied directly. If you dig in raw or untreated products, though, add nitrogen and wait for about a month before planting.

For unplanted areas, green manuring (planting a crop for the sole purpose of digging it under to enrich the soil) provides one of the best and cheapest sources of organic matter. Popular green manures include legumes (fava beans, soybeans, clover, vetch, and alfalfa, for example) and other plants such as kale, mustard, and rye. Legumes have the advantage of adding nitrogen to the soil. Whatever crop you choose, dig it under while still green—legumes just as they begin to flower, other crops before flowering.

Be aware that adding organic matter won't magically change your soil overnight. Improvement takes time, especially if you're starting out with loose sand or rock-hard clay. Add organic matter frequently, at least yearly; you can't do it just once or twice and expect the benefits to last forever. When organic material decomposes, it's converted to nutrients that are absorbed by plants—after which you need to enrich the soil again. The depletion process occurs quickly in hot climates, more slowly in cooler regions.

How much to add? For the most marked improvement, a good rule of thumb is to incorporate a 3-inch layer of organic matter into the upper 8 to 9 inches of soil. Spread the material over the surface, then thoroughly mix it in. Never work the soil when it's wet; wait until it's moist, but easily crumbled.

Working it in. Rototilling is fine for establishing a garden bed, but you're better off using a garden fork to dig in regular additions of organic matter. Routine use of a rototiller is counterproductive, since it crushes the

A pH meter allows you to measure the relative acidity or alkalinity of your soil.

soil and interferes with soil organisms. It can also create a compacted area (at the depth of the tiller's tines) that roots and water have difficulty penetrating.

Soil pH

Measured on a scale from 0 to 14, pH refers to relative acidity or alkalinity. The neutral point is 7; a pH value lower than 7 is acidic, while one higher than 7 is alkaline (sometimes called basic). Acid soils are common in high-rainfall areas, alkaline soils where rainfall is low.

A well-amended soil usually has a pH between 6 and 6.8, the range in which the vast majority of plants thrive. Many woodland plants, such as azaleas, rhododendrons, and blueberries, need a more acidic soil (pH 4 to 5.5); many desert plants prefer a more alkaline one (pH 7.5 to 9).

Most plants can't extract nutrients from soils with a pH outside their preferred range. Iron, zinc, and manganese become less available with increased alkalinity, while nitrogen, phosphorus, and magnesium are among the nutrients less available as acidity rises. Plants suffering from nutrient deficiencies exhibit symptoms that are often mistaken for pest or disease damage; see page 13.

To learn your soil's pH, send a soil sample to a laboratory for testing. Or do the test yourself; most of the pH meters and kits sold by garden suppliers are reasonably accurate. If you discover that the pH is seriously out of whack for the plants you want to grow, you'll need to correct the problem. (Consult a good gardening encyclopedia for information on the pH requirements of specific plants.) Soil sulfur raises acidity; ground limestone makes soil more alkaline. The report from a professional lab will tell you how much material to add; you can also check with your Cooperative Extension office for recommended amounts.

MAKING COMPOST

You can use a variety of methods to turn yard and kitchen waste into the dark, humus-rich soil amendment and mulch known as compost. Techniques vary from simple to painstaking; the degree of participation is up to you. At one extreme, you can just let a pile of debris sit and rot for a couple of years. At the other end of the spectrum, you can carefully construct a pile that heats up rapidly, yielding finished compost in as little as a few weeks.

Compost is created by billions of bacteria, fungi, and other organisms which consume organic matter and leave the undigestible part behind as compost. To do their work, they need carbon, nitrogen, moisture, and air. The closer you get to the ideal proportions of these elements, the more efficiently the compost-makers will feed—and the sooner you'll get compost.

A well-managed pile should contain roughly equal volumes of nitrogen-rich ingredients (fresh or green materials such as grass clippings, manure, and vegetable peelings) and carbon-rich ingredients (dry or brown materials such as dried leaves, twigs, straw, and shredded paper). Keep out diseased and pest-ridden plants, unless you're confident that the pile will get hot enough—131° to 140°F is the optimum range—to kill the organisms. (To check the temperature, stick a composting thermometer into the middle of the pile.)

For efficient decomposition, the pile should be at least 3 by 3 by 3 feet; a bin isn't necessary, but it neatens the composting area.

Classic composting calls for alternate layers of carbon-rich materials, nitrogen-rich materials, and soil, but soil isn't really needed—and neither are neat layers. It's actually better to mix the materials. Chop up large pieces: the smaller the chunks, the faster they'll decompose. When you add food waste, cover it with sawdust or old compost to keep scavengers away.

The heap should be about as moist as a damp sponge. Plenty of air is essential; otherwise, bacteria that operate in the absence of oxygen will take over the pile and make it reek. Turning the compost frequently is the traditional way to supply air, but there are other methods. One of the simplest techniques is to set a roll of chicken wire or a wide-diameter PVC plastic pipe with holes drilled along its length upright in the bin, then pile the waste around it. This method is even more effective if you set a wooden pallet at the bottom of the heap, so that air can flow up from beneath.

There's no evidence that adding fertilizers, lime, or other amendments is useful. Commercial compost starter isn't necessary either, since organic matter contains all the microorganisms required for efficient composting.

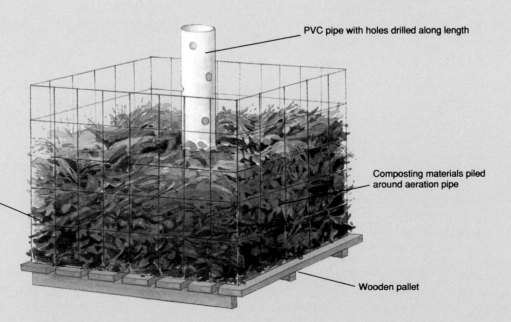

PVC pipe with holes drilled along length

Minimum 3' x 3' x 3' pile for efficient decomposition

Composting materials piled around aeration pipe

Wooden pallet

Silvery aluminum-foil mulch reflects the blue sky, confusing some flying insects and discouraging them from landing on nearby plants.

Mulching

Mulch—a protective layer of material spread over the soil surface—contributes to good plant growth in many ways. It conserves moisture, insulates roots against temperature extremes, keeps the soil from crusting over, and suppresses weeds. Mulch also prevents rainfall and irrigation water from splashing disease organisms up from the soil surface onto plants.

Trees planted in a lawn will benefit from an encircling ring of mulch. The mulch not only reduces competition from the surrounding turf, but also protects trees from two of their worst enemies—lawn mowers and string trimmers. The wounds inflicted by these machines offer disease organisms easy entry.

Mulches act as a barrier to some pests. A thick layer of hay or straw on potato plants slows down adult Colorado potato beetles as they dig their way to the surface after pupating in the soil; it also conceals the plants, making it difficult for the beetles to find dinner. A mulch layer prevents potato tuberworm moths from crawling through soil cracks to lay eggs on tubers. A sheet of plastic laid under a leaf miner–infested plant catches the larvae and keeps them from entering the soil to pupate.

Colored Mulches

Plastic mulches in various colors are a new trend in pest control. Reflective aluminized plastic, which mirrors the blue sky and confuses some flying insects (winged aphids and thrips, for example) is widely available. Aluminum isn't the best choice for heat-loving crops like melons, though, since it reflects the sun's heat. Some garden suppliers carry black plastic blended with gray; it mimics the silvery aluminum, but absorbs heat as well.

For a do-it-yourself reflective mulch, place strips of aluminum foil on either side of transplants or just-planted seeds. Or spread out a single sheet of foil and plant through slits cut in its surface.

Studies have shown that red repels root maggot and other flies, while orange repulses sweetpotato whiteflies. These results may not hold true outside the testing areas, however; scientists caution that pests in different regions may react differently. Degradable colored mulches are currently being developed; watch for news of them.

Organic Mulches

The advantage of an organic mulch—such as ground bark, homemade compost, or leaf mold—is that it improves the soil as it breaks down. Many gardeners dig in the mulch after each growing season, but even if the material is left on the surface, it will gradually break down and work its way into the soil.

Most organic mulches should be laid 2 to 4 inches thick, although very light materials can be applied up to 6 inches thick. Don't pile the mulch against woody stems: it can rot the plant and attract pests that gnaw on bark.

It's true that some organic mulches, especially coarse-textured types with lots of hiding places, encourage slugs, snails, and other pests that appreciate moist, dark, cool living quarters. On the positive side, such mulches also attract beneficial creatures like earthworms and spiders. To entice garden allies, the mulch should be applied during planting and replenished each season—or during the season, if it breaks down rapidly. (See pages 35–36 for more about welcoming beneficial creatures.)

PLANTING FOR SUCCESS

*W*hether you're aiming for a flourishing flower garden or a bumper crop of vegetables, several factors are involved in a successful planting. The first step is choosing the right plants—those well adapted to your growing conditions and (if possible) resistant to local troublemakers. Location and timing—where and when you plant during a particular year—are important as well. The most decisive factor, though, is proper planting procedure. Unless plants are installed correctly, the best stock and the cleverest planting strategies will be wasted.

Choosing the Right Plants

"The right plant in the right place" is the rule that guides successful gardeners. Plants stressed by an inhospitable climate or unsuitable growing conditions suffer from stunted growth, poor color, or malformations; they're also more vulnerable to pests and diseases than are their well-adapted counterparts.

Stroll or drive around your neighborhood and notice what grows well. Find out more by talking to local gardeners about their successes and failures. Visit a nearby nursery or contact your Cooperative Extension office for information on the plants best suited to your area.

Don't stint on plants. Always opt for the highest quality in seeds and planting stock. Choose resistant plants (see below) and certified disease-free stock when they're available. Deal with reputable suppliers so you can be reasonably certain that the material you buy is healthy and vigorous. If you're unsure about the condition of plants you've obtained from another gardener, isolate them for a few weeks to let any problems come to light.

Select healthy-looking plants with good leaf color. Don't automatically take the biggest plants you see; they may seem like the best bargain, but compact specimens are actually healthier than sparse leggy ones. A mass of blooms is appealing, but plants loaded with buds will transplant better and flower longer than those which have already blossomed. When buying trees, compare the trunks: the larger the diameter, the sturdier the tree. Also remember that the trunk should taper upward from a thick base, not go straight up like a telephone pole.

Check plants carefully and reject any showing signs of pests or diseases. Also refuse any with wounds, which weaken plants and make them disease-prone. A container plant should be well-rooted but not rootbound—that is, it shouldn't have large, circling roots or a mass of roots growing from the bottom of the container. A plant that's obviously too big for its pot is also likely to have crowded roots.

Although starting off with large plants can give you an instant landscape, there's a lot to be said for smaller transplants. Not only are they more affordable, but they get established more quickly—and they'll catch up to and often outgrow larger transplants within a few years.

Resistant Plants

You may be able to avoid battling certain pests or diseases if you can find plants that resist the trouble-

The hybrid musk rose 'Ballerina' is resistant to black spot, a fungal disease of roses common in rainy climates.

makers. In some cases, an entire plant type is resistant; for example, there's a long list of plants that resist verticillium wilt, among them grasses, conifers, oaks, willows, dogwoods, pears, nasturtiums, and zinnias. If you know that your soil is infested with the fungus, you'll save yourself trouble by favoring plants that aren't susceptible.

In other cases, only certain varieties of a plant are resistant. Tight-husked types of corn thwart entry by corn earworms, for instance. Old-fashioned roses are generally far less disease-prone than hybrid teas and other modern kinds, although some of the modern types do resist certain diseases.

Some plants are resistant by nature, while others get their resistance from cross-breeding. Until recently, resistant strains could only be developed by continually reproducing plants showing the desired characteristics, but today scientists are focusing on gene-splicing as a means of producing resistant types.

The most common form of resistance involves defense mechanisms providing protection from disease. The plant may contain a chemical that inhibits development of the infection; or perhaps it lacks a nutrient the disease organism needs. Insect-resistant plants exist, too, but they're far less available than disease-resistant types. Insect resistance may result from a taste that repels the insect, a growth habit that hampers attack, or simply a good tolerance for feeding by the pest, with no ill effects on yield or appearance.

Resistance is sometimes indicated on the plant label. For example, the letters VFTN after the name of a tomato variety indicate resistance to verticillium wilt,

Choosing plants adapted to the growing conditions is one of the home gardener's best assurances of success. These sun-loving flowers are ideally suited to their location.

fusarium wilt, tobacco mosaic virus, and nematodes. You may also see PM (for powdery mildew–resistant) and DM (for downy mildew–resistant) on some labels. Usually, however, you'll have to rely on catalog companies or gardening books to tell you which plants are resistant to which problems.

Because pests and diseases may exist in several strains, it's a good idea to consult your Cooperative Extension agent or a nurseryman for information on specific plant varieties resistant to local pest types. In most states, the Cooperative Extension Service publishes lists of good performers that defy pests and diseases prevalent in the region.

Keep in mind that resistance is often temporary; it may be overcome by a particular pest, or a more aggressive strain of the troublemaker may develop. To keep one step ahead of the game, breeders are constantly developing new and improved varieties.

Of course, growing resistant plants doesn't mean you can neglect other prevention techniques: even if a plant's resistance is still effective, there's no guarantee it will work for you during a given year. The plant may still succumb if conditions are stacked against it.

Using Planting Strategies

Timed plantings and crop rotation are two devices gardeners commonly use to thwart some pests and diseases.

Timed Plantings

If certain pests show up in your garden every year around the same time, you may be able to dodge serious damage by planting their favorite crops earlier or later than usual. Think of it as eating early or late to avoid feeding uninvited guests who always drop by during the dinner hour.

Early planting can save corn from corn earworms and protect gladiolus from thrips. Set beans out in advance of your usual schedule to foil spider mites, which are most troublesome when the weather turns hot; by the time the pests show up, the plants will be sturdier and better able to resist attack.

Sometimes insects appear in greater numbers and feed more heavily early in the season; in this case, a delayed planting will deprive them of their favorite food. A large part of the population may die out or move on to better pickings before getting interested in your crops.

Another strategy is to plant early in hopes that pests *will* attack the planting. This type of timed planting, called a trap crop, is effective against such creatures as Mexican bean beetles and squash vine borers. You set out a few plants before the usual planting date, as bait to draw the pests away from a crop you actually intend to harvest. The trap crop needn't be the same plant as your harvest crop; it can be anything else the pest relishes. Planting in containers is fine if you don't want to waste garden space. In any case, wait until the trap planting is infested; then dispose of it, pests and all.

Keep a record of your planting dates, so you can make adjustments from season to season. Don't ignore the local climate, though—you'll only harm plants if you subject them to overly cold or hot weather in an effort to evade pests.

Crop Rotation

If you don't grow your crops in the same place year after year, you can duck certain pests and diseases, especially those that overwinter in the soil. Obviously, crop rotation isn't practical for permanent plantings such as rose gardens and fruit orchards, but it's an easy way to prevent problems in annual flower and vegetable beds.

COMPANION PLANTINGS

Many gardeners firmly believe that pairing certain plants will discourage pest attacks. For example, growing garlic or chives with roses is supposed to keep Japanese beetles away. Besides onion-family plants, the most commonly used protective companions are pungent herbs and some flowers—notably nasturtiums, marigolds, chrysanthemums, geraniums, and petunias.

Unfortunately, there's not much scientific support for the value of companion plantings. Much of the information is based on intuition and folklore. You'll find just as many gardeners extolling the virtues of a particular pair as declaring it a total bust.

The most frequently reported success story involves the use of marigolds to thwart pest nema-todes. In the past, gardeners were advised to grow solid stands of French dwarf or African marigolds, but recent evidence has revealed that interplanting the flowers also helps. Other pairings, however, have shown poorer results. The problem is that you have to plant the protective companion so densely that you wind up reducing the yield of the crop you're trying to protect.

It can't do any harm to try pairing plants and see what happens. If you want to put some teeth into your conclusions, run a controlled experiment: grow paired plants in one area and raise the susceptible plant alone elsewhere. Use at least two groups of several plants each, and run the test for several seasons before reaching a verdict.

Crop rotation works by interrupting the life cycle of the harmful organism. The idea is that if you deprive pests of the only food they'll eat, they'll die out. Without tomato-family plants to infest, for example, potato tuberworms will starve to death. Likewise, onion maggots will succumb if they can't find onions. And a soilborne fungus such as downy mildew will die if it doesn't have a susceptible plant to infect.

Since closely related plants are often vulnerable to the same pests and diseases, they should not succeed each other for several years in a row. Most gardening experts suggest a 4- to 6-year rotation, but greater intervals may be necessary to discourage certain fungi that can survive for even longer periods without a host.

Rotation is most often practiced in vegetable gardens. By growing the following related crops together and moving them as a group each year, you'll avoid a buildup of destructive organisms. (*Be sure* to rotate the tomato family, since its members are notoriously prone to soilborne diseases that prosper when the hosts go into the same ground every year.)

■ Tomato family: Tomato, potato, pepper, eggplant
■ Onion family: Onion, shallot, leek, chive, garlic
■ Beet family: Beet, Swiss chard, spinach
■ Cole crops: Cabbage, cauliflower, broccoli, Brussels sprouts, bok choy, collards, kale, kohlrabi, mustard, radish, rutabaga, turnip
■ Legume family: Bean, pea, cowpea, peanut
■ Carrot family: Carrot, celery, celeriac, parsley, parsnip
■ Cucurbits: Cucumber, melon, pumpkin, squash
■ Lettuce family: Lettuce, chicory, endive

Planting Properly

Before you plant, be sure you've picked a suitable spot—one that satisfies the plant's requirements for light, temperature, soil, and various other factors (humidity and shelter from wind, for example). Also make sure to provide adequate air circulation around plants, since some pests and diseases flourish in crowded, overgrown conditions.

Extreme temperatures stress new plantings, so avoid planting just before or during the hottest or coldest times of year. Fall planting is preferable in mild-winter regions, spring planting in cold-weather climates (though fall planting in chilly regions is fine for many hardy species). In all areas, bare-root plants should be installed during the dormant season.

If you have a choice, plant on a calm, cool, overcast day that will be followed by a week or so of the same sort of weather. If you must plant when conditions are less favorable, be prepared to shelter the newly installed plants for a few weeks or until they produce new growth. A row cover (see page 48) will shield them not only from hot sun and harsh weather, but also from pests. Other devices for short-term protection of small plants include overturned strawberry baskets, paper bags, and paper cups.

Improve your chances of raising sturdy, healthy plants by using the following planting techniques.

Seeds

Seeds can be sown directly in the ground or started in containers and moved to the garden later. Either way, your biggest concern will be damping-off, a fungal disease that stops seeds from germinating or kills seedlings after they emerge. Take preventive measures as explained on page 99.

Follow the seed-packet directions for recommended planting dates and spacing. The optimum planting depth will also be noted; as a rule of thumb, you should scatter tiny seeds, then press them in gently, and sow larger seeds no deeper than three to four times their diameter. When direct-seeding in the garden, sow twice as many seeds as you'll need, since some will surely be lost to pests and unfavorable weather.

Don't sow unless you can provide adequate warmth: seeds can rot in cold soil that stays wet. Be sure to label your plantings, since many seedlings look alike when they first emerge. Mist to keep the soil moist. When seedlings planted indoors have germinated, move them into bright light to keep them from becoming leggy.

Once true leaves (not just initial seed leaves) have developed, thin seedlings to avoid overcrowding and to provide good air circulation. Don't pull up the extra seedlings; just pinch or snip them off at soil level. Also apply a dilute liquid fertilizer, unless you mixed in a time-release fertilizer at planting time.

A week or two before indoor seedlings are ready to go into the garden, accustom them to the outdoors (a process called "hardening off") by taking them outside each day for periods of gradually increasing length. Start by placing the seedlings in a shady spot for a few hours, then expose them to more sun bit by bit.

Annuals & Perennials

To keep transplants from drying out, water them well and remove them from their containers only after digging the planting holes. Each hole should be about twice as wide as and no deeper than the rootball.

If you're setting out seedlings from flats, cut them apart. If there's a mat of interwoven roots at the bottom of the rootball, tear it off; otherwise, just loosen roots by pulling apart the bottom third of the rootball. Set the plant in the hole, making sure the top of the rootball is even with the surrounding soil. Fill in with soil, firming it around the roots; then water well.

Trees & Shrubs

The standard advice for planting trees and shrubs has changed in recent years. Experts used to recommend digging large holes and amending the backfill soil with lots of organic matter, but research has shown that most plants with large root systems grow better when planted in native soil.

Whether you're installing container, bare-root, or balled-and-burlapped plants, dig the hole the same way (see below). In each case, you'll form a ridge of soil around the planting to serve as a watering basin.

Remove a container plant from its can and uncoil circling or twisted roots. Set the plant in the hole and spread the roots. Fill in around the roots with un-amended soil until the hole is about half full, firming the soil with your fingers. Water thoroughly to remove air pockets, finish filling the hole, and water again. The top of the rootball should remain slightly above grade.

To install bare-root plants, build a cone of soil in the middle of the hole and spread the roots over it. The plant should be at the same depth as it was in the growing field; a change in color on the bark shows where the soil line was. Fill in soil and firm as you would for a container plant.

Slide a balled-and-burlapped plant into the hole, making sure that the top of the soil ball is slightly above grade. If the wrapping is genuine burlap, untie it and

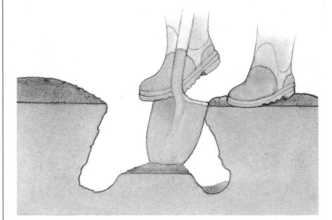

All tree and shrub planting begins with a hole. Dig it so that sides taper outward into the soil and are roughened, not smoothly sculpted; this lets roots penetrate more easily into surrounding soil. To prevent or minimize settling of the plant after planting and watering, make the hole a bit shallower than the rootball or root system of the plant it will receive; then dig deeper around edges of the hole's bottom. This leaves a firm "plateau" of undug soil to support plant at proper depth.

spread it to uncover about half the rootball, but leave it in place to disintegrate. If the "burlap" is a synthetic material, however, remove it to allow the roots to grow into the surrounding soil. Fill the hole, but if the soil in the rootball is substantially different from the surrounding soil, add some organic matter to make a gradual transition. Also punch holes in the rootball to improve water penetration.

KEEPING YOUR GARDEN HEALTHY

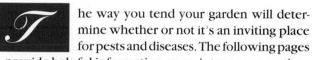

T he way you tend your garden will determine whether or not it's an inviting place for pests and diseases. The following pages provide helpful information on maintenance practices designed to keep troublemakers away.

Maintaining Vigor

A dependable source of water and nutrients helps keep plants healthy and equips them for self-defense against pests and diseases.

Watering

Plants differ in their moisture needs. Some grow better in moist soil, while others prefer the soil to dry out between waterings. Some types tolerate drought in the shade but need water in the sun; others don't drink much no matter where they're located. If you don't know your plants' moisture requirements, ask a nursery professional or consult a plant encyclopedia. And remember—all newly installed plants, including drought-tolerant ones, need regular watering until they're established.

Soil type will strongly influence your watering schedule. Sandy soils require frequent attention, since water gushes right through them. Clay, on the other hand, absorbs water slowly and drains slowly too, so it shouldn't be watered too often or at too rapid a rate. You'll have fewer watering worries if you improve the soil with organic matter (see pages 22 and 24), since soils with a high organic content are both fast-draining and moisture-retentive.

When you water, wet the entire root zone. If you just sprinkle the top inch or so of soil, you'll encourage roots to stay near the surface, where they'll dry out quickly. Experience will teach you how much and how often to water. A few hours after watering, dig down to

A cold frame warms the spring soil, allowing an earlier start for warm-weather crops such as corn. The plastic covering offers the added advantage of barring pests.

see how deeply moisture has penetrated; if it hasn't reached the plant roots, leave the water running longer. Be careful not to give *too* much water, though: overwatering produces lots of succulent new growth, which attracts aphids and other pests. Excess water also encourages disease organisms such as water-mold root rot fungi, which thrive when water stands around roots.

The way you apply water can affect plant health. Overhead watering wets leaves, making the plant more prone to disease. If you use sprinklers, run them early in the day, so the foliage has time to dry by nightfall. Drip irrigation, practical for all plantings except lawns, keeps leaves dry and thus decreases their susceptibility to diseases.

Finally, keep in mind that water should be applied when the plant needs it, not according to a preset schedule. Your garden will need more water when the weather is hot, sunny, dry, or windy than when conditions are cool, overcast, humid, or calm.

Fertilizing

The three major nutrients all plants need are nitrogen, phosphorus, and potassium; ten other nutrients are required in smaller quantities. Nitrogen, essential for shoot growth, is the element that must be replenished most often: it's in short supply in all soils, since it's used by plants in large amounts and washes out of the soil easily. Keep in mind, however, that too much of a good thing can cause problems. Excess nitrogen produces foliage at the expense of flowers and fruit; it also encourages a flush of succulent growth that draws pests and increases susceptibility to fungal diseases.

MAINTAINING A HEALTHY LAWN

If you follow the guidelines outlined on these pages, you should have no trouble maintaining a flourishing lawn—as long as the variety of grass you grow is adapted to local conditions. An unfit lawn is prone to all sorts of problems, so consider replacing it with a more suitable grass if possible.

Mowing

Most homeowners mow on a set schedule, but grass doesn't grow according to a timetable. If you mow too often when the lawn is growing slowly, you may damage the roots and encourage pests and diseases. It's just as bad to let grass get too tall before cutting. The key is to mow regularly enough to keep the lawn at its proper height (ask your nurseryman what's ideal for your grass variety), but to remove no more than a third of the grass blade at each mowing. If your lawn's ideal height is 2 inches, for example, mow when it reaches 3 inches.

By leaving the clippings on the lawn to decompose (the lawn industry calls this "grasscycling"),

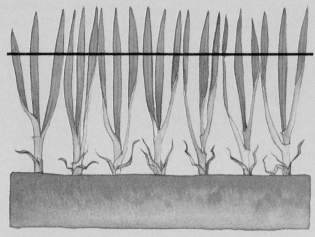

Mow so that you remove no more than one-third of the blade at a single mowing.

you can cut down on watering and fertilizing. Grass blades are about 95 percent water, so they'll disappear within a week or so, adding organic matter and

Phosphorus is essential for healthy root growth, fruit and seed formation, and early maturity. Since this element doesn't leach from the soil, you don't need to add it very often. Highly acid soils are most likely to be phosphorus-deficient.

Potassium contributes to general vigor and disease resistance. Though plentiful in most western soils, it's often in insufficient supply in the East. Like phosphorus, potassium doesn't easily leach from the soil, so it doesn't require frequent replacement.

Unless their organic content is extremely low, most soils contain the other minerals needed for plant growth in adequate quantity. Iron, zinc, and manganese deficiencies do frequently occur in the West, though, since alkaline soils make these elements unavailable to plants. Adding the minerals in chelated form (which plants can absorb more easily) or acidifying the soil usually solves the problem.

Fertilizers vary in their nutrient content and the form in which they're applied; they also differ in how rapidly they release nutrients. Some fertilizers are complete (containing all three major nutrients); others are simple (providing only one major nutrient). The various fertilizer forms include granules, pellets, liquids, and powders. Some products are soluble and release their nutrients quickly, while others release them slowly over time.

By law, fertilizer labels must list the percentages by weight of nitrogen, phosphorus, and potassium—in that order. For example, a 12-6-8 fertilizer contains 12 percent nitrogen, 6 percent phosphorus, and 8 percent potassium; the remainder is filler. If one of the major nutrients is entirely lacking, a zero will appear on the label; a 0-10-10 fertilizer contains no nitrogen, for example.

Some products are labeled for special purposes; you may find "rose food" or "camellia food," for instance. Such fertilizers can be used on a variety of plants, but they're particularly suitable for the type mentioned.

Instead of using packaged fertilizers, some gardeners rely on regular additions of organic matter. Manure is one choice; it must be aged before being dug in, then watered well to leach out the salts. The potential nutrients to the soil. They'll break down even faster if you use a mulching mower, which shreds the clippings before depositing them on the lawn.

Keep mower blades sharp. Dull blades tear grass instead of cutting it, making the lawn more susceptible to disease.

Watering

Every time you water, the grass should be soaked to the roots. To find out how long you must water to reach the desired depth, use a soil probe or dig down. If the water runs off instead of soaking in, cycle the irrigation: turn off the water as soon as it starts to run off, wait for about 10 minutes, and then turn it back on.

Even in dry climates, it shouldn't be necessary to water a lawn more than twice a week. Many water companies, especially those located in the West, offer guides telling you how much and how often to water.

Fertilizing

When grass clippings are routinely left on the lawn to decompose, very little supplemental fertilizer is needed. When you do fertilize, use a type that releases nitrogen slowly. Quick-release nitrogen spurs growth, which means more frequent mowing. Overfertilizing can also lead to a thirsty lawn, thatch buildup, and diseases.

Aerating

Aerate the lawn (poke holes into it) once or twice a season to enhance the absorption of water and nutrients. If you rent an aerator from a tool shop, ask for the kind that removes cores of soil instead of just punching holes.

Dethatching

When thatch (a thick mat of grass stems and roots at the soil surface) exceeds a thickness of ½ inch, it blocks water and nutrients and should be removed. Some people think thatch is due to grass clippings left on the lawn, but it's actually caused by improper watering, too much fertilizer, and soil compaction.

You can physically remove thatch with a vertical mower (which cuts out chunks of thatch) or a de-thatching rake, but topdressing is gentler. Apply a thin layer of organic matter or sand over the lawn; as the material filters down, it encourages the growth of microbes that slowly decompose the thatch.

drawback of relying solely on organic materials is that they may not release enough nutrients at once to give plants what they need for optimum growth. Organic matter must be broken down into nutrients by soil-dwelling organisms, so it's generally effective as a fertilizer only when the soil is moist and warm enough for these organisms to be active.

Protecting Plants

For gardening success, choose plants that are able to withstand the extremes of your particular climate. Remember, too, that a plant is better able to cope with temperatures somewhat higher or lower than usual when it's installed in a location that suits its needs. Get to know all the little microclimates in your garden: the spots that are warmer or colder, sunnier or shadier, more humid or drier, and windier or calmer than the other areas.

Early in the season, row covers and hot caps (covers for individual plants) will keep seedlings and young transplants warm and shield them from pests as well (see "Coverings," page 48). To give young trees the protection they often need from intense sunlight, wrap tree-wrap paper or burlap around the trunk or whitewash the bark.

If you live in a harsh-winter climate, you may have to cover some plants to protect them from the cold. Laying pine boughs or straw over an entire bed will insulate roots, although these materials can also harbor mice. Find out what's necessary in your area; see what your neighbors do. The goal is not to warm up the plants, but rather to stabilize them, since alternate freezing and thawing causes plant cells to rupture and decay. It may be necessary to relegate very tender plants to containers and move them indoors when the weather turns chilly.

In mild-winter climates, even light frosts can cause havoc. If you have advance notice, you may be able to save tender plants by covering them with burlap or plastic (secure the cover over frames or stakes, so it doesn't touch the plants). Another option is to mist plants frequently to keep the tissues from freezing.

Get rid of plant debris, which can provide breeding sites or winter living quarters for pests and disease organisms.

Thinning & Weeding

Getting rid of excess or undesirable growth neatens a garden, but that's not the only reason for doing it. Thinning overgrown shrubs and trees (removing selected branches right down to the base) opens the plants to sunlight and improves air circulation, thus discouraging many diseases. Don't overdo it, however: plants suffer when they don't have enough foliage for photosynthesis, the chemical process by which they manufacture energy to fuel their growth. If the plant is susceptible to bark beetles or borers, don't prune when these insects are active.

Weeding promotes a healthy garden in at least two ways. First, because weeds compete for water, nutrients, light, and space, removing them gives garden plants a chance to grow more vigorously. Left to fight it out unaided, your desirables will usually take a poor second place, showing defeat in stunted growth, fewer flowers, and reduced harvests. Second, since some pests and diseases spend part of their lives on weeds, you may be able to disrupt their life cycles by eliminating these plants.

It's true that some weeds shelter beneficials, but you're still better off growing special plants to attract helpful creatures (see pages 35–36) and wiping out the weeds.

Keeping the Garden Clean

When their favorite hosts aren't available, many insect pests and disease organisms survive on plant debris. By cleaning up the garden, you can eliminate breeding and overwintering sites.

Throughout the growing season, get rid of dead, diseased, or infested vegetation; don't let it pile up near plantings. Shred prunings that may contain borers. When infested fruit drops prematurely, you may be able to interrupt the pests' life cycle by burying it. (Of course, if a pest can survive burial, you'll have to get rid of the debris some other way.) At the end of the season, organize a cleanup of the whole garden.

Put disease-ridden growth in a compost heap only if you're confident that the pile will attain high enough

USING REPELLENTS

Gardeners employ many materials to repel pests. Some of the most common types are intended to keep larger creatures away—for example, bags of mothballs are used to deter raccoons and deer. Other deer repellents abound, including bars of deodorant soap and mesh bags filled with human hair or blood meal. The trick is to keep changing the repellent before the animals get used to it.

Healthy, insect-free leaves or bulbs, blended with water and sprayed on crops susceptible to attack, are a widely used type of insect repellent. Natural gardening books are replete with homemade spray recipes, usually calling for onions, garlic, hot peppers, and pungent herbs. Various oils, such as lavender, cedar, and pennyroyal, are also used as repellents. When sprayed on plants, brews made from water and the wood chips or sawdust of certain plants—cedar and quassia, for example—are supposed to rebuff pests as well.

Many gardeners add a little dishwashing liquid or vegetable oil to a spray to make it stick. If you concoct your own spray, always try it on a small part of the plant first; wait for a day or so, then check for damage before coating the entire plant. Reapply sprays every few days, and be sure to renew them after a rainfall.

temperatures (around 140°F) to kill the organisms. Some persistent weeds (experience will tell you which ones) can survive even hot composting, so dispose of them in another manner.

Because many disease organisms are easily spread by infected soil clinging to tools and shoes, it's important to clean up thoroughly after working among diseased plants. To sterilize tools, dip them in a solution of nine parts water to one part bleach or rubbing alcohol.

Inspecting the Garden Regularly

If you're always on the lookout for problems, you can eradicate them before they establish a foothold. Survey the garden for pests and diseases at least weekly, more often when you know there's trouble brewing.

When you make your inspection, take along a 10-power hand lens, a pair of pruners or scissors, plastic bags for collecting samples, and a notebook for recording your findings. Be prepared to handpick any pests you discover (handpicking techniques are described on pages 43–44).

As you walk through the garden, check every few plants for signs of infestation or actual pests. Turn over leaves and check nooks and crannies. Look for damaged or discolored foliage and insect droppings. To examine a plant thoroughly, follow the procedure detailed on page 16.

Learn the habits of pests, so you know their hiding places and dining hours; then inspect at their preferred mealtimes and try to catch them red-handed. If the pests are nocturnal, schedule nighttime sorties or look for the troublemakers in their daytime hideaways. If you've set traps, such as overturned flowerpots for catching snails or rolled-up newspapers for snaring earwigs, check them daily and dispose of your catch. Otherwise, all you're doing is providing free shelter.

Welcoming Beneficial Creatures

Wherever you find pests, you'll usually find natural enemies that attack them. Since these allies are better pest-control agents than the average gardener can possibly hope to be, it only makes sense to nurture and protect them. You can even go a step further and purchase beneficials (see pages 50–56), but those that arrive on their own are more likely to stay. Moreover, a garden that attracts natural enemies is less apt to have pest problems calling for the importation of helpful creatures.

The most obvious way to protect garden allies is to

A hand lens provides a close-up view of trouble during garden inspection tours.

avoid using controls that are harmful to them. One of the materials that spares beneficials is the pathogen *Bt*, various strains of which target caterpillars, certain beetles, and some mosquito and fly larvae. Among pesticides, botanical types generally present less danger than do synthetics, since they break down faster. The botanical pesticide ryania is safe for beneficials for another reason as well—the toxin must be eaten along with the treated foliage, but beneficials don't feed on plants. (For more information about pesticides, see the next chapter.)

When using a product that may harm garden allies, don't cover the entire area on the same day; wait for the material to dissipate from one section before treating another. If a pesticide is potentially injurious to bees, apply it in the evening, when bees are less active.

Beneficial creatures will reproduce in gardens that fulfill their basic needs. Since most adult beneficials are tiny and have mouthparts that can't reach deep into blossoms, accommodate them by growing small, shallow-necked flowers. Among the best choices are members of the carrot family (*Apiaceae*, also called *Umbelliferae*) and the sunflower family (*Asteraceae*, also called *Compositae*). Carrot-family plants include carrot, dill, celery, parsley, coriander, fennel, anise, and parsnip. The sunflower family is a huge group of about 20,000 species, among them sunflower, chrysanthemum, daisy, marigold, dahlia, yarrow, and zinnia.

Sunflower-family members don't produce as much nectar as carrot-family plants do, but they bloom over a longer period. Many other types of flowers also entice various beneficials, so grow an assortment and see which creatures show up. If you're short of space, try scattering a wildflower seed mix developed especially for your area.

To persuade beneficial creatures to stick around, give them hideaways. Crevices formed by rocks or pieces of wood make a cozy home for ground beetles, lizards, and many other garden helpers. Of course, a protective environment may also appeal to pests, so you'll have to weigh the pros and cons of offering hidey-holes. For example, a coarse-textured mulch with lots of crannies attracts spiders, but it may also draw slugs and other undesirable elements.

Insects find the lure of sex irresistible. Gardeners living in parts of the East where spined soldier bugs are found can attract these creatures by setting out cones containing a pheromone scent. In the future, this type of lure may be available for other beneficial insects. (See pages 47 and 56 for more about pheromones and spined soldier bugs.)

No matter how appealing the garden, beneficial creatures will move on unless there are enough pests to satisfy them. You can help keep some allies around by not pulling away the plate before they've finished dinner. For example, if you see ladybird beetles or lacewing larvae around an aphid-infested plant, leave that plant alone.

Of course, you'll have to learn to distinguish partners from pests. The following pages contain descriptions of some common garden allies; local gardening experts can tell you about other helpers that may frequent your area. For descriptions of pests, see the rogues' gallery beginning on page 71.

BENEFICIAL CREATURES

BEES

Honeybees (shown), bumblebees, and many native bees are the primary pollinators in a garden. Throughout their lives, they rely entirely on nectar and pollen for nourishment. As they sip nectar, pollen clings to their hairy bodies, then is dispersed when they alight on other flowers. Honeybees also feed pollen to their young, collecting it in "pollen baskets" on their hind legs. Like wasps and sawflies (their close relatives), bees have two pairs of wings.

Bees are extremely susceptible to many pesticides, including botanical types. To protect these insects, use materials that degrade quickly and apply them late in the evening, when bees are in their hives. Never apply long-lasting insecticides to flowers in bloom.

CENTIPEDES

These brown, flattened, fast-moving creatures hide in dark, damp places (under debris, logs, or stones, for example) during the day, then emerge at night to feed on insects, slugs, and snails. They're equipped with a pair of venomous claws, which they use to seize and paralyze their victims.

Despite their name, centipedes don't have an even 100 legs; the number ranges from 30 to several hundred. (There's one pair of legs for each body segment, so the total leg count depends on how long the creature grows.) The centipedes found in American gardens are usually around an inch in length, but in the tropics they may reach a foot long.

EARTHWORMS

Earthworms don't attack pests, but they're indispensable in the garden. They feed on organic matter, which they digest and excrete as castings, an excellent nitrogen source for plants. Their greatest contribution, though, is their tunneling activity. Earthworm tunnels aerate the soil, improve drainage, and serve as channels for root growth. Earthworms also help build good soil by pulling decomposed organic matter from the surface down into the ground.

GROUND BEETLES

Ranging from pea-sized to over an inch long, most ground beetles are shiny black or blue-black, although some have a metallic sheen of green (or another color) on their wing covers. Both adults and larvae feed voraciously on cutworms, root maggots, and other soft-bodied pests found at or below the soil surface. Some types consume slugs and snails.

Most ground beetles are night feeders; during the day, they hide in the soil or under debris. Calosoma beetles (shown), also known as fiery searchers, secrete an irritant; if you handle them, wear a pair of gloves.

LACEWINGS

The handsome adults (shown), with transparent wings jutting out beyond their bodies, feed on nectar, pollen, and honeydew. The voracious, alligatorlike larvae (see page 4) pursue aphids, spider mites, and other soft-bodied pests, grasping their prey with pincerlike jaws and sucking them dry.

Green lacewings can be purchased (see page 53), but both green and brown species occur naturally in gardens. Adult green lacewings reach about ¾ inch long, their larvae around ½ inch; brown lacewings are slightly smaller. Brown species deposit white eggs directly on plant surfaces, while green types lay tiny green eggs singly atop slender filaments cemented to leaves and stems.

LADYBIRD BEETLES

Commonly called ladybugs, these insects include hundreds of species that vary widely in color. Some are red with black spots; others come in various shades of orange, brown, or yellow; still others are solid black. A few types can be purchased, primarily the convergent ladybird beetle and the mealybug destroyer (see pages 52 and 53). Most species feed on aphids and other soft-bodied insects, spider mites, and insect eggs.

Female beetles typically lay clusters of elongated orange eggs that stand on end. These hatch into tiny alligatorlike larvae (bottom photo); they're usually blue-gray and orange in color, with raised black spots. The larvae devour more prey than the adults (top photo), which also consume nectar, pollen, and honeydew. Both adults and larvae are usually found feeding together where prey is plentiful.

PARASITIC WASPS

A diverse group of insects, parasitic wasps are generally tiny and host-specific. The females have an ovipositor—a tube at the rear end that's used to deposit eggs, singly or in masses. Parasitic wasps lay their eggs on or inside a target pest; the young wasp typically develops within its host, although some species just pupate inside the victim or on its skin. Cocoons attached to the backs of hornworms (as shown) and hollow aphid bodies with a neat exit hole are two familiar signs that wasps are on the job.

Braconid, ichneumon, and chalcid wasps are among the most common parasitic types.

Braconids are small wasps with transparent wings and long antennae. The adults of most species are black, although some are yellow or bright red. Target pests include hornworms, gypsy moth caterpillars, tent caterpillars, cutworms, and aphids.

Ichneumon wasps make up a huge family; they're slender, with long antennae similar to those of braconids. They vary greatly in size, with the ovipositor adding inches in length to the females of some species. Most adults are black, brown, yellow, or red. Ichneumons lay eggs in many types of caterpillars, including corn earworms, armyworms, and codling moth caterpillars.

Chalcids encompass several families of minuscule wasps whose favored victims include aphids, whiteflies, leafhoppers, caterpillars, beetles, and scale insects. Most adults are black, although some species are blue, green, or yellow. Two types of chalcids commonly sold (and discussed in "Parasitic Wasps," pages 54–55) are trichogramma wasps and whitefly parasites.

PRAYING MANTISES

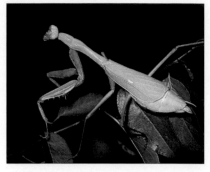

Up to 5 inches long, these fascinating creatures derive their name from the way they hold their large, spiny forelegs folded together, as if in prayer. (In fact, the legs are positioned to seize and hold prey.)

Although praying mantises are sold as predators (see page 55), don't expect much pest-control assistance from them: they're indiscriminate feeders, devouring just about anything that crosses their path. The insects overwinter in egg cases attached to plant stems or house siding; they hatch in early summer and reach adulthood in early fall.

PREDACEOUS BUGS

Certain true bugs prey on soft-bodied pests, piercing the villains' bodies and sucking the fluids. Some of these predators have quite colorful names: assassin bugs (shown), ambush bugs, pirate bugs, big-eyed bugs, damsel bugs, and wheelbugs. Spined soldier bugs (see page 56) are one type sold for pest control.

The adults of most species are about ½ inch long and extremely agile. All are shield-shaped (sometimes quite elon-gated) and bear a triangular marking on the thorax. Both adults and nymphs are predatory, dining on aphids, leafhoppers, small caterpillars, thrips, and spider mites.

PREDATORY WASPS

Various types of wasps hunt other insects (primarily caterpillars) and spiders. A predatory wasp stings its prey to paralyze or kill it, then carries the body back to its nest or burrow. Social species such as paper wasps share the catch with the rest of the colony; solitary types like mud daubers stuff their victims directly into the partitioned cells housing wasp larvae. Both social and solitary wasps defend their nests, but the former are quicker to sting humans when provoked.

ROBBER FLIES

This group of flies earns its name from the adults' habit of lying in wait for prey, then darting out with a loud buzzing noise and snatching victims in midair. Robber flies are far from selective feeders, ambushing all kinds of flying insects. The ¼- to 1¼-inch-long adults vary in appearance—some resemble flies, others small bumblebees, and yet others wasps—but all have bulging eyes, long legs, hairy bodies, and a long proboscis that's used to stab prey.

Female flies lay white eggs in the soil. These hatch into tiny, white, flattened maggots, which live underground and feed on grubs and grasshopper eggs. Taking up to 2 years to complete their life cycle, robber flies overwinter in the soil as larvae.

ROVE BEETLES

Although the many species of these usually nocturnal beetles vary in size and shape, all can be recognized by their short wing covers. The abdomen is almost completely exposed and often curls upward when the beetle runs. The hind wings (beneath the abbreviated wing covers) can be unfolded quickly for immediate flight.

Rove beetles are found wherever there's decaying organic matter, but not much is known about the habits of most species. Many types are predatory, feeding on such pests as slugs, snails, and root maggots; some kinds parasitize the pupae of other insects.

SOLDIER BEETLES

Slender, ½-inch-long adult soldier beetles are usually red or orange, with brown, black, or gray wing covers. Often seen on flowers, the beetles dine mainly on pollen and nectar, but also consume small insects. The grubs, which hatch from clusters of eggs laid in the soil, live underground and feed on soil-dwelling pests.

SPIDERS

Indiscriminate in their appetites, spiders catch prey either by hunting or trapping. Hunters actively pursue their quarry and inject them with venom; they're more aggressive than trappers, which weave webs and wait for victims to fly into them. Jumping (shown), lynx, crab, and wolf spiders are among the most efficient hunters.

Of the few spiders that pose danger to humans, the most poisonous are the black widow (a shiny black spider with a red hourglass on the abdomen) and the brown recluse or violin spider (matte brown, with a darker brown, violin-shaped mark on the head).

SYRPHID FLIES

Both adults and larvae are about ½ inch long, and both help gardeners: the nectar-feeding adults pollinate flowers, while the larvae prey on pests.

The adults are called "hover flies" for their helicopterlike flying ability. Their black- and yellow-banded bodies give them the look of bees or wasps, but they don't sting and have only one pair of wings instead of two. The larvae—small, green or light brown, and wormlike—use fanglike hooks to hoist their victims and drain them of fluids. Favored prey includes aphids (the flies often lay their elongated white eggs in aphid colonies), mealybugs, and other small insects.

TACHINID FLIES

There are more than a thousand species of these flies. Resembling large houseflies, the bristle-covered adults are usually gray or brown with pale markings. Nectar feeders, they're most often seen perching on flowers or buzzing around vegetation near the ground.

The tiny, greenish white, spined maggots primarily parasitize caterpillars, but also afflict beetles, grasshoppers, true bugs, and other pests. They enter their victims in various ways. Typically, female flies cement their white eggs to the host, usually on a caterpillar's head; once the eggs hatch, the maggots eat their way inside. Some flies lay large numbers of eggs on plants; pests eat the eggs, which then hatch within their bodies. Other species inject incubated eggs into the host or let hatched maggots find their own hosts to infiltrate. In all cases, the maggots develop within the pests and eventually kill them. When they're done feeding, they usually leave the host and pupate on the ground.

VERTEBRATE PREDATORS

Various birds, including flycatchers, swallows, warblers, and nuthatches, are allies in the battle against bugs: they consume huge numbers of insects.

Other insect-eating predators, such as garter snakes (top photo), salamanders, lizards (bottom photo), toads, and turtles, often arouse animosity—but their appearance or method of locomotion shouldn't diminish your appreciation of the valuable services they perform. They eat all kinds of insects, and they're less likely than birds to catch agile beneficials.

Preventing Problems   **39**

Beneficial lacewing larvae, shipped in corrugated cardboard cells, scramble over camellia leaves in search of prey.

DEALING WITH PESTS & DISEASES

After you've identified a pest problem, you're ready to deal with it. That may mean treating it—or deciding to leave it alone. Some so-called problems don't amount to much, especially since perfect gardens exist only in the dreams of gardeners. Learning to live with a few chewed or spotted leaves is much easier (and allows more time for actual gardening) than struggling against nature. Moreover, the pesky insects so many of us decry provide food for other, more beneficial creatures. There are good reasons for tolerating other pests, as well: who can imagine a garden without birds, even if some of them are irksome at times? And for many, the sheer pleasure of hosting butterflies is well worth the sacrifice of a few plants to hungry caterpillars.

Of course, if a problem is truly serious, it should be tackled vigorously. There's no sense in allowing pests to run riot if you can do something about them.

Once it's clear that action is warranted, you must decide how to proceed. Begin by learning all you can about the offending pest or disease; the rogues' gallery starting on page 71 will help you. Next, choose your weapons. Pesticides are a familiar solution to pest problems, but such chemical remedies aren't the only controls available. The astute gardener can also choose from physical and biological controls.

Physical controls, such as handpicking and trapping, are the most direct. Although these tactics may sometimes be impractical for large commercial operations, they're highly effective in home gardens.

Biological controls allow you to encourage a naturally occurring process—one in which beneficial creatures keep harmful ones in check. Among the helpful organisms available to gardeners, some are insects that prey on or parasitize pests; others are pathogens (disease-causing microbes like *Bt*) that infect pests.

Many gardeners are inclined to avoid *chemical controls* entirely—but these remedies are still an important component of IPM, although they should be used judiciously and as a last resort. The chemical arsenal includes not only synthetic pesticides, but also mineral-based products, soaps, oils, and botanical pesticides.

Always evaluate pesticide use. If a chemical is ineffective after the first application, think twice about using it again for the same problem. And if other methods give you results as good as or better than those achieved by chemical means, stick with the nonchemical remedies instead.

Remember, too, that growing a healthy garden (see "Preventing Problems," pages 21–39) can largely relieve you of the necessity of dealing with pests.

TAKING ACTION

basic maxim of thoughtful pest management is to act against pests only when they're causing intolerable damage. Obviously, what constitutes "intolerable" depends on the individual. The discovery of a few nibbled leaves or the odd wormy apple drives some gardeners to immediate action, while others may not do anything until whole plantings wither and die. Wise gardeners occupy the middle ground: they learn to tolerate small afflictions but are prepared to battle potentially big problems—especially those that, unchecked, will kill valuable plants or ruin entire harvests.

Deciding When to Act

Commercial growers who follow IPM principles use "economic injury level" as the threshold for action: when the predicted cost of a crop loss exceeds the cost of pest control, it's time to act. Of course, a grower must often consider a crop "lost" as soon as it's blemished, since the public is largely unwilling to accept less-than-perfect produce. Gardeners, however, can afford to be much more forgiving of minor damage.

When you do decide that the damage in your garden can't be ignored, the next step is to make sure you're pinning the blame on the right pest. Pages 7–19 offer tips for recognizing trouble and identifying the probable culprit. Also refer to the charts on pages 106–110 for a list of some common plants and the pests that typically attack them; then check the pest descriptions starting on page 72 to see if you can single out the guilty party. If you're still not sure what's causing the trouble, look for help at a reputable nursery or contact your local Cooperative Extension office.

Even when you've identified the problem, don't spring into action until you've determined whether the situation is being resolved on its own. Has nature intervened on your behalf? Perhaps the offender is succumbing to predators or unfavorable weather conditions. On the other hand, if the pest population is growing, you'll need to take a hand in the matter.

In some cases, you'll lose your crop if you put off treatment until damage is evident. Nonetheless, it's best not to take preemptive action until you're reasonably sure how much injury a pest might cause. Sticky traps (see page 45) and pheromone traps (see page 47) will tell you if certain insects are reproducing in large enough numbers to cause significant harm. As soon as you know that a problem is developing, waste no time in implementing controls.

If a problem is so bad that you fight a losing battle every year, you may be better off not growing the susceptible plant. Don't look on this decision as throwing in the towel; it's a perfectly legitimate means of pest control.

Choosing Controls

When you're ready to select particular control measures from among the many choices available to you, keep in mind that a combination of methods usually works best: in pest control, as in most other matters, it doesn't pay to put all your eggs in one basket. If you implement just one measure and it fails, you may worsen the problem. You might also introduce new problems; for example, if you rely exclusively on a broad-spectrum insecticide, you might kill the beneficial insects keeping other pests in check.

The choices you make will depend on a number of factors. The most important of these is efficiency: will the method work against the pest? More specifically, will it work against the life-cycle stage you intend to battle? Insects that undergo complete metamorphosis (see "How Insects Grow," page 8) change so drastically as they develop that different remedies may be appropriate at different stages. If you're to succeed, the pest you're targeting must be at a stage susceptible to the control: there's no sense in spraying the caterpillar-killing microbe *Bt* on adult moths or applying a protectant fungicide to a plant that's already badly infected, for example.

Pheromone traps, such as this codling moth trap in a pear tree, will tell you if certain pests are present in large enough numbers to warrant action.

Safety is another very important consideration. Among workable controls, always opt for those presenting the least hazard to people, pets, wildlife, and the environment.

Next, consider practicality. Handpicking large, slow-moving pests is a fine means of control if only a small area is infested, but not if an entire large garden is involved. And no matter what the extent of infestation, it's a waste of time trying to grab pests that are barely visible or that dart away when you approach.

Take the probable permanence of the cure into account, as well. Handpicking Japanese beetle adults is effective, but treating the grubs with milky spore disease (see page 57) is more likely to produce lasting results.

Think about ease of use, too. If you want to spray a towering shade tree but don't own the equipment you need to reach the top, you may decide to hire a professional to do the job—which brings up another factor, cost. If a problem is getting out of hand and spraying is the best solution, hiring someone with a powerful pump sprayer may be worthwhile. But if you detect only a few caterpillars on your tomato plants, crushing them is easier and cheaper than spraying, even if you have the right equipment.

Keep personal preferences in mind, since you're more likely to implement controls that appeal to you—or, at the very least, don't repel or intimidate you. For instance, you may prefer to release beneficial insects or use row covers rather than handle chemicals.

Be aware that some control methods take time, so you may not see results for weeks or even for an entire season. In the case of using milky spore disease on Japanese beetle grubs, the wait may be as long as 3 years.

Available controls fall into several categories. You can manage pests with physical tactics, pit beneficial creatures and disease-causing organisms against them, or apply an assortment of chemicals. Each of these categories includes a number of techniques, each effective against certain pest problems. Consider the pros and cons of each method as regards your specific situation.

PHYSICAL CONTROLS

 hysical controls are your first line of garden defense. These simple, time-honored tactics require no pesticides, but instead deal with pests strictly by mechanical means: handpicking and trapping, erecting obstacles, or using scare tactics. Garden suppliers sell some very imaginative equipment, from inflatable owls and snakes to more aggressive devices such as spiked sandals: just strap them on and march around the lawn, impaling grubs as you aerate the grass.

By themselves, mechanical controls may not solve a problem, but they often reduce it to a tolerable level. Sometimes, though, you'll need to supplement physical tactics with other types of controls.

Handpicking

This is the easiest, most direct way to get rid of clearly visible, slow-moving pests. Take along a 10-power hand lens to ensure correct identification, since you don't want to kill beneficial creatures by mistake (see pages 36–39 for descriptions of beneficials).

It's best to handpick in the early morning, when most insects are relatively sluggish. If you're hunting nocturnal pests such as snails, schedule a nighttime foray and bring a flashlight.

Careful observation is the key to a successful handpicking campaign. When inspecting plants, remember to scan both the upper and lower leaf surfaces

Jittery pests like asparagus beetles are easy to knock into a container of soapy water.

and the stems. Some pests, such as imported cabbage-worms, blend in visually with foliage, so look with a keen eye. Also check each marauder's known haunts and hiding places, such as the top few inches of soil or under debris. Handpick pests at any and every stage you come across: eggs, larvae, pupae, and adults. You won't get every single pest, but you will reduce the population.

Different techniques work for different pests. To kill aphids, run infested plant surfaces between your fingers. Pick relatively large insects like caterpillars individually, then crush them or drop them into soapy or oily water. When you're dealing with skittish types like asparagus beetles, it's most efficient to place a container of soapy water beneath your quarry—the pests will tumble in when you reach for them. You can shake Mexican bean beetles by the dozens out of plants onto a cloth spread on the ground. Japanese beetles play dead when disturbed, so they too can be shaken from their roosts onto a cloth, then disposed of. Hand-pick leaf miners, insects that tunnel between upper and lower leaf surfaces, by removing infested leaves. A long-handled pole or broom will help you obliterate webs sheltering tree caterpillars. To kill tree borers, ram a piece of wire into holes where you see sap exuding. To get a squash vine borer, make a lengthwise slit in the stem and pluck the borer out.

If employed fairly early in the game, handpicking is also effective in checking the spread of some diseases: just pick off afflicted leaves before they can infect the rest of the plant. If you remove mildewed leaves soon enough, you can slow the disease. (If the infection is caused by a virus, however, handpicking won't help. Since these cases can't be cured, pull up the whole plant to keep the virus from spreading.)

Vacuuming

Commercial growers use huge agricultural vacuum cleaners to suck pests out of their crops. You can do the same with a household machine. Good candidates for vacuuming include squash bugs, whiteflies, spider mites, and other pests that congregate in large numbers and obligingly remain in the vicinity to be suctioned up, rather than scattering when you draw near. Catch them early in the morning, when they're at their most lethargic, and don't forget to vacuum both leaf surfaces.

A hand-held wet/dry vacuum is the safest device, but you can use a regular model if you keep it on dry ground and use a ground fault circuit interrupter. Small battery-operated models are also safe to use, but they may not provide sufficient suction. Gardeners report that both crevice and brush attachments are effective.

To make sure that pests don't escape, encase the vacuum bag in plastic and freeze it overnight; then throw it in the trash.

Spraying Water

A blast from a garden hose is a simple control that reaps benefits far outweighing the time and effort required. The stream of water dislodges or injures many pests, such as aphids, mealybugs, lacebugs, and spittlebugs. You do need a good jet of water to knock pests from their perches, but don't make it so powerful that you damage the plant. Be sure to hit the undersides of leaves, where many pests are found.

Hosing down a plant also creates an unfavorable environment for pests such as spider mites, which prefer dusty, dirty plants and low humidity. To avoid creating conditions favorable for diseases, spray early in the day, so that leaves have time to dry before nightfall.

Pruning

Sometimes you can get rid of pests simply by pruning off the infested part of the plant. When aphids cluster on tender shoot tips, pinch off and discard the growth. By removing tree branch tips harboring fall webworms, you eliminate pests and unsightly webs at the same

Pruning off infested or damaged growth is often the simplest solution to a pest or disease problem.

time; prune back to a fork so that you don't leave stubs or stimulate a lot of weak new growth. Of course, snipping away problems is impractical if you're dealing with tall trees or extensive infestations.

Pruning is successful in keeping some diseases from spreading. For example, control measures for the bacterial disease fireblight always include pruning off blackened shoots as soon as you see them. (For more about pruning fireblighted twigs, see page 100.)

Tilling

Spading the soil and turning it over kills some soil-dwelling pests and exposes others to unfavorable weather and predators; still others will be buried so deep they won't be able to crawl to the surface.

Don't wait for spring to till: if you know that a pest spends the winter below ground, turn the soil over in autumn. Always leave some areas untilled to provide refuge for predaceous ground beetles and other helpful creatures living in the soil. Permanent, mulched pathways between beds serve this purpose nicely.

Using Traps

Some pests can be nabbed with simple homemade traps. Certain undesirables will readily hide under objects placed on the ground near their favorite plants: snails crawl beneath overturned flowerpots, earwigs take shelter in rolled-up newspapers, and squash bugs snuggle under boards or pieces of burlap. Jars filled with a solution of sugar or molasses in water attract flying insects, which then fall in and drown. Sunk into the ground, jars containing a water-molasses mixture become grasshopper traps. To make another simple trap, push carrots or pieces of potato partway into the soil to attract wireworms, then throw out the infested pieces. Lure harlequin bugs away from crops by placing cabbage leaves elsewhere in the garden. In all these cases, be sure to collect and destroy captured pests daily; otherwise, you're only providing them with a cozy home.

Though there's some debate about how well they work, beer traps are often recommended for enticing slugs and snails to a watery grave. Keeping the traps filled with beer can be expensive—but as it turns out, you don't really need beer. It's not the alcohol that draws the pests, but rather the fermented combination of sugar, water, and yeast. To brew your own bait, mix ⅜ teaspoon active lager yeast (available at home-brewing supply stores) or baking yeast and 1 tablespoon sugar with each cup of water. Pour the liquid into a

Snails will seek shelter beneath inverted flowerpots, propped up slightly to allow the pests easy access. Turn the pots over and dispose of your catch daily.

shallow pan and set it in the garden; pests attracted to it will fall in and perish. Extra solution can be stored in the refrigerator for several months. In the garden, it should be changed weekly.

Garden suppliers sell a variety of devices designed to snare specific pests. Some traps are for killing, others simply for monitoring, and yet others for capturing pests for later release. Many work by enticing the target creature with food, shelter, or sex. Sex lures consist of a chemical attractant that mimics the mating signals of particular insects; see "Pheromone Traps" (page 47). Other common traps are discussed below.

Sticky Traps

These traps use colors to lure insects to their doom. The pests don't see colors as we do, but rather respond to the wavelength of light a particular color emits; heading towards the source of that wavelength, they collide with the trap and stick to its adhesive coating.

Many insects (including adult whiteflies, winged aphids, fruit flies, and thrips) are attracted to yellow, which they seem to perceive as a patch of bright foliage. Some thrips—especially those that infest blue flowers—are also drawn to light blue. Flea beetles, tarnished plant bugs, and rose chafers are attracted to white, apparently mistaking it for white buds or flow-

This yellow sticky trap is simply a piece of stiff cardboard coated with a semigloss alkyd paint. A commercial sticky material was applied to the dry surface.

ers. Because insects vary in their response, some experimentation may be needed to find a material emitting the right wavelength for the target species.

Sometimes, the shape of the trap is as important as the color. For example, red spheres hung in apple trees draw apple maggot flies, while green spheres placed in walnut and peach trees lure walnut husk flies. (For more on fruit flies, see page 80.)

Sticky traps can be used for monitoring pests or for mass trapping. To increase the chances of success, commercial traps often incorporate an attractant either in the adhesive or attached to the trap. For example, you can buy red spheres with lures derived from apple odors.

Though sticky traps are widely sold, you can make your own from paper cups, pieces of plywood or stiff cardboard, or other objects painted the appropriate color and coated with an adhesive. Apply semigloss alkyd paint to all surfaces to make the trap waterproof. For yellow sticky traps, use a bright, sunny-yellow paint. To make red spheres, paint old croquet balls or burned-out light bulbs fire-engine red; for green spheres, use dark green paint. When the paint is dry, brush on

a sticky coating. The commercial sticky goo available through garden suppliers usually remains sticky even after a downpour.

Place sticky traps at foliage height, either suspended from branches or set atop stakes. Space the traps 10 to 15 feet apart for monitoring, only a few feet apart for mass trapping. If you're hanging traps in trees, place one in each dwarf tree, two to four in a medium-size tree, and six or more in a large tree.

Light Traps

Once widely used by commercial growers to monitor pests, these devices (usually ultraviolet lamps) have been largely supplanted by pheromone traps. However, some garden suppliers still offer light traps, including models intended to electrocute insects.

Light traps have several drawbacks. Since they operate only at night, pests that do damage during the day are unaffected. Because the light attracts all kinds of nocturnal insects—not just the bad ones—beneficials may be killed along with harmful creatures. Finally, pests may escape, since only those insects flying *into* the trap actually die.

Animal Traps

Animal traps include both live and killing traps. If you're not sure whether trapping a particular animal is lawful, check with your state game or conservation department.

Live trapping is not always as benevolent as it might seem. One problem is finding a new location for the captured animal. Some states prohibit the release of certain animals; and even when legality isn't an issue, you're just passing your problem on to your neighbors if you let the animal go in an inhabited area. In some cases, animals cannot adapt to their new homes. And the relocation of certain creatures, such as raccoons and squirrels, carries the danger of spreading diseases. (The SPCA will help you trap some kinds of animals; for information, contact your local office.)

No permission is needed to trap and kill some pests, most notably voles, pocket gophers, and moles. Mole and gopher traps are among the most commonly sold killing traps—and, in fact, trapping is the best way to get rid of these wily creatures (see pages 94 and 95 for details).

Whatever type of trapping you do, get the proper trap and bait for the animal you want to catch, and be sure to follow the manufacturer's instructions. Handle all materials with gloves so you don't get your scent on them. In general, traps should be placed in animal runways or between the animal's shelter and the plants

PHEROMONE TRAPS

Insects produce *pheromones*, chemicals that regulate the behavior of other members of the same species. Many of these very specific scents have been duplicated in the laboratory. At least one such synthetic pheromone is used to attract a beneficial insect—the spined soldier bug (see page 56)—but most are intended for pest control. Laboratory-produced pheromones exist for dozens of pests, including black cutworms, cabbage loopers, codling moths, corn earworms, and peachtree borers. (Attractants derived from fruit or floral odors are available for trapping apple maggot flies and Japanese beetles, but these aren't pheromones.)

As a means of pest control, pheromones (usually sex pheromones) are used in several ways: to capture insects, as lures to monitor the density of a pest population, and as signals to disrupt mating.

The key component of a pheromone trap is a rubber or plastic lure impregnated with or encasing the attractant. The lure is placed in the trap, where it slowly releases a come-hither signal. Traps come in various shapes and sizes and may be made of plastic, paper, or other materials; most have a sticky coating or a funnel-shaped opening, so that insects are caught after entering and cannot escape. (Spined soldier bug lures aren't enclosed in a trap. The attractant comes in a yellow cone that you hang on a branch or push into the ground; it draws both male and female bugs.)

Capturing insect pests. Using sex pheromones for mass trapping is chancy, since only males are attracted; if some of them slip through and mate with females, you'll still have a problem, though it may be less severe. Codling moths are among the pests often targeted for mass trapping, but success isn't guaranteed by any means. You'll need enough traps to handle all the pests, and each trap must work flawlessly. Neighborhood cooperation is important. Otherwise, so many pests may arrive in your garden that your traps are overwhelmed, or mated females from neighboring yards may fly in to lay eggs. On the other hand, mass trapping may do the trick if the pest population is low anyhow; or it may at least reduce the population to acceptable levels.

Monitoring pests. Pheromone traps are usually best used for monitoring pest populations: when a certain number of insects are trapped within a specific time frame, you know the population is high enough to warrant action on your part. To find out the threshold figures, ask the pheromone manufacturer or supplier or consult your Cooperative Extension office. These sources can also advise you on the number, placement, and timing of traps.

Disrupting mating. In addition to their use in traps, pheromones are now employed to disrupt mating. In an area saturated with the attractant, male insects can't locate females: either their sensors overload or they embark on a wild goose chase. "Disrupting" attractants are available for several pests, including codling moths. The dispensers, which look like twist-ties, work only when large areas are treated.

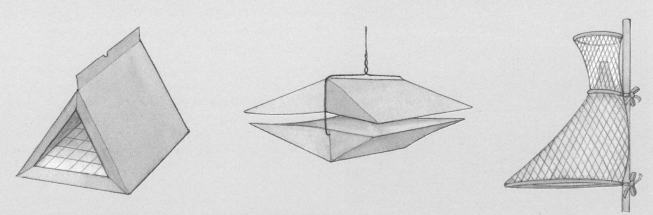

Pheromone traps come in various configurations. Common types include (from left to right) the delta trap, wing trap, and heliothis trap. The supplier will tell you which type is appropriate for the insect you want to trap.

it prefers to eat. Make the traps look natural by covering them with soil or grass; check daily to see if you've caught anything.

Erecting Barriers

A physical barrier set up between pest and plants is often the easiest means of eliminating pest damage. The barricade can be as simple as a bottomless orange juice can or paper cup placed over a seedling to deter cutworms, or as elaborate as an electric fence to stop deer short. Steep furrows around beds guard successfully against armyworms, which fall into the trenches and can't crawl out again. A wire-mesh lining under a raised vegetable box keeps out pocket gophers. The five types of barriers described below are widely used in home gardens.

Coverings

Row covers, cages, and netting placed over young plants keep many pests at bay. Although some of these

Protect strawberries from birds and other fruit-stealing animals by growing the plants in a cage made from chicken wire and PVC pipe.

products were developed for frost protection or other purposes, their ability to thwart destructive garden visitors has made them popular for pest control.

With the edges sealed, row covers present an impenetrable barrier to pests migrating into the garden, especially tough-to-control kinds like whiteflies and cucumber beetles. The covers prevent flying insects from laying eggs on or near plants; they also protect plants from larger pests such as rabbits and deer.

To control pests that overwinter in the soil, such as root maggots and hornworms, combine row covers with crop rotation (see page 28–29) and tilling (see page 45). Check regularly for pests beneath the covers; if they're left undisturbed and unmenaced by their natural enemies, they can multiply quickly.

Some row covers are designed to fit over frames or hoops, while others rest directly on the plants. The latter type, called a floating row cover, is made of spun fiber and is extremely lightweight and porous; it lets in air, water, and light while excluding flying insects. For maximum protection, apply floating row covers before seedlings emerge and bury the edges in the soil; be sure to leave enough slack to allow for plant growth.

Remove row covers when plants are sturdy enough to fend for themselves; or, if necessary, take the covers off to allow pollination. Row covers that retain heat should be removed in summer—if you leave them in place, you may cook your garden.

If your goal is to keep birds from pecking at tender foliage, a fine-mesh cage over seedlings will do the trick. Wire cages made from rolls of mesh or old window screens also frustrate deer and many other animals.

Netting over fruit trees and berry bushes will stymie fruit-stealing birds. If the plants you're guarding bear fruit at the branch tips (as do blueberries, for example), build a frame that holds the netting about a foot away from the plant. Otherwise, birds will peck the fruit through the mesh or knock it to the ground.

Copper Barriers

Copper barriers around garden beds actually keep out slugs and snails: the combination of slime and copper apparently produces either an unpleasant chemical reaction or a slight electrical charge. In any case, the mollusks refuse to cross copper. Just be sure you don't leave any gaps or foliage bridges that could offer the pests access to your plantings.

Many garden suppliers sell 3- or 4-inch-wide copper strips by the foot; you can also buy copper sheets and cut them up as needed. Band the strips around tree trunks or tack them to the frames of raised beds. Other copper products include copper foil (sold in rolls) and

The most common barriers include (from left to right) paper-cup collars on seedlings to foil cutworms; commercial sticky material painted on woody stems to stymie climbing pests; and mesh cylinders wrapped around tree trunks to deter gnawing animals.

a sticky, copper sulfate-containing material that you paint on stems and trunks. (See below for more on sticky barriers.)

Collars

The collar most often seen in home gardens is a bottomless tin can or paper cup, slipped over seedlings and twisted an inch or so into the ground to keep cutworms from chewing off stems.

The honeycombed interior of a corrugated cardboard band wrapped around a tree trunk provides an ideal place for codling moth caterpillars to pupate. You can make your own cardboard collars or buy them; remember to destroy the cocoons weekly.

Root maggot flies don't like to lay eggs on tar paper—hence the success of tar paper collars around susceptible seedlings. Simply cut a square or circle of tar paper, then make a slit in the center to accommodate the plant stem.

Squash vine borer moths usually lay eggs on cucurbit stems near the soil; to discourage them, foil-wrap that part of the stem.

Sticky Barriers

Designed to thwart climbing pests, the sticky goo sold at garden supply stores can be sprayed, painted, or troweled onto stems and trunks. The gummy barricade deters pests such as ants and root weevils from crawling up into plants to feed. Sticky products based on vegetable gum are less likely to harm plants than petroleum-based materials are, but both types must be renewed frequently. (These products are messy, so keep children and pets away from them.)

To protect young trees, wind plastic wrap around the trunks, then paint the sticky material on that. You can also buy a tree wrap with adhesive coating already in place. To make a similar wrap, fasten a strip of cotton batting about 2 inches wide around the tree trunk; cover it with a piece of tar paper, then apply the sticky coating. Instead of the standard goo, you can use roofing tar.

Fences

Fencing is the surest way to bar animals such as deer, rabbits, raccoons, and woodchucks from your garden, especially in areas where these creatures flourish. You can fence the entire garden or just cordon off individual plants. The best type of fence for the job depends on the animal; see the pest listings on pages 93–97 for more information.

To deter animals that can climb or jump ordinary fences, you may decide to resort to electricity. An electric fence jolts an animal, but doesn't kill it. Some suppliers sell battery-powered types you can erect permanently or take down after each year's harvest.

Taller fences are for keeping out deer, lower ones for banning raccoons and other small animals. (Because electric fencing has proved ineffective against deer in some regions, investigate its record of success in your area before you buy.)

Using Scare Tactics

While pests are supposedly driven off by frightening images and sounds, there isn't much substance to most such claims. A scare tactic may or may not succeed, depending on how it's used and how skittish the pests are; it may work on pests in one garden and not in another, or it may be effective only briefly. The scare tactic may even get the credit for driving out pests when in fact other factors are responsible.

There's little evidence that scarecrows, inflatable snakes and owls, or dangling pie tins keep birds and other wildlife away. However, you may improve your chances by changing the "frighteners" periodically or moving them around the garden before animals become accustomed to their presence in certain sites.

Some suppliers sell kites in the likeness of predatory birds such as hawks or buzzards. Since a kite swoops in the wind, it's more apt than a stationary object to scare off birds and other small animals.

Noisemakers abound, ranging from plastic pinwheels to ultrasonic gadgets that emit high-frequency sounds or vibrations. Some of these devices are touted as surefire ways to banish burrowing beasts—but in view of the fact that pocket gophers and moles often tunnel along railway tracks and airport runways, the claims are doubtless exaggerated.

Leaving a radio on all night in the garden is often recommended as a means of expelling raccoons; playing recordings of alarmed birds is a standard suggestion for scaring birds away. Such tactics may work, but they usually provoke a worse problem—the wrath of your neighbors.

BIOLOGICAL CONTROLS

 hen gardeners deploy natural enemies that prey on or parasitize pests, or when they release pathogens (disease-causing microbes) that infect and kill pests, they're using biological controls.

Pitting helpful organisms against destructive ones may seem like current science fiction, but it's actually

Birds soon become accustomed to fixed objects—but they may be spooked by these evil-eye balloons, which are rigged on pulleys to move around the garden.

an ancient practice. As long ago as 300 A.D., the Chinese established colonies of predatory ants in citrus orchards to control caterpillars and tree-boring beetles. Modern use of biological controls in the United States dates back to 1889, when the U.S. Department of Agriculture (USDA) imported the vedalia beetle from Australia to battle the cottonycushion scale, a pest imperiling the California citrus industry. The insect was subdued within months and has not been a threat since.

The use of natural enemies grew steadily until the 1940s, when the development of "miracle" pesticides such as DDT pushed biological controls into the background. The prevailing belief was that the powerful new synthetic chemicals would solve pest problems so easily and inexpensively that all other controls would be rendered obsolete. By the 1960s, however, it was obvious that these marvels of modern science weren't so marvelous after all. Many products were harmful to both people and the environment; and by killing helpful organisms and pests indiscriminately, they upset the balance of nature, often creating worse problems than they solved.

Today's search for less toxic means of pest control has brought biological tactics back into the spotlight. Since many pests are immigrants from other parts of the world, scientists have concentrated on finding and importing the organisms that keep each pest in check in its homeland. As a result, more and more biological agents are available to commercial growers and home gardeners every year.

Predators & Parasitoids

Pests' natural enemies are either predators (creatures that feed on pests) or parasitoids (organisms that live on or inside a pest, ultimately causing it to die). You'll often find parasitoids referred to as parasites, but technically speaking the two terms aren't synonymous: parasitoids kill their hosts, while true parasites coexist with the host, usually weakening it but rarely ending its life. (To confuse matters even further, the adjective "parasitic" is often used to describe parasitoids; the parasitic wasps discussed on pages 37 and 54–55 are in fact parasitoids.)

Most predators eat a variety of prey, although some prefer a particular target. Parasitoids, in contrast, usually depend on a single host or a narrow range of hosts. The majority of parasitoids are wasps so tiny that most people are unaware of their presence.

Using natural enemies offers numerous advantages. The creatures are nontoxic. Since they don't eat plants, they don't become garden pests. Their numbers

rise and fall in concert with the size of the quarry population. When pests increase in number, so do beneficials, since more prey is available; when the pest population shrinks, beneficials starve or disperse in search of other prey. Consequently, a healthy garden maintains a balance between harmful creatures and helpful ones.

You can "seed" an area with a small number of beneficials in hopes of establishing a population that will suppress pests over the long term. Or, if you have a specific pest problem, you can inundate the afflicted site with natural enemies to overwhelm the pests.

If an insect or mite infestation is really serious, you may want to apply a pesticide before you try biological controls. The pesticide will reduce the pest population somewhat, ensuring that the beneficials you later release will have a fighting chance. If you go this route, use a nonresidual material, such as a soap or oil, or a selective pesticide, which kills a narrow range of pests. (For information on chemical controls, see the section beginning on page 58.) Before releasing the natural enemies, wait until the pesticide has broken down into harmless substances. (It's hard to know when this point has been reached; as a guide, use the days-to-harvest or days-to-reentry interval listed on the pesticide label.)

Don't expect beneficials to wipe out a pest population completely. As the pest's numbers dwindle, natural enemies become less efficient at finding the last few survivors. And when the number of pests falls too low, the enemies will simply disappear from your garden.

Even with large-scale releases, natural enemies won't provide overnight relief: it takes time for the creatures to do their work. For this reason, it's important that you release beneficials *before* a crop is in imminent danger. If you know from past experience that an insect against which you can deploy a natural enemy ravages your garden every year, schedule the release when the pest is present but hasn't yet built up to damaging numbers.

Biological agents are usually raised in insectaries or laboratories and sold through mail-order sources (see page 69). Since most are lightweight and compact, they're inexpensive to ship. Some garden centers sell parasitic wasps and other perishable creatures—sort of. What you actually get is a package containing a prepaid envelope that you dispatch to the supplier when you're ready for the shipment. Time your order so that you can release your beneficials shortly after you receive them. Release should coincide with the pest's most vulnerable stage: if a natural enemy feeds on eggs, for example, it should be deployed just as egg laying begins.

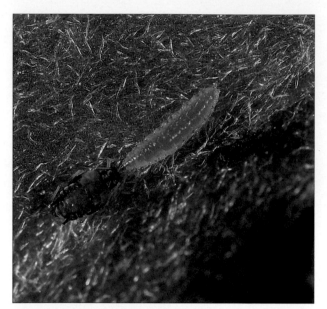

This tiny orange aphid midge larva will suck the body fluids from its aphid victims.

If you've never released natural enemies before, you may want to begin with green lacewings and spined soldier bugs, both of which prey on a wide variety of pests. Some suppliers offer variety packs containing assorted predators and parasitoids.

Your shipment will come with instructions for release. The release rates cited in the following descriptions are guidelines; discuss your needs with the supplier. Repeated releases are often required.

There's no guarantee that natural enemies will stay in your garden long enough to do their job. Releasing them in the relative cool of the evening may improve your chances. If you're deploying six-legged helpers, you might try inducing them to remain by distributing a commercial insect food throughout the garden. Beneficials are also more likely to stick around if your garden provides refuge for them; for ways to keep your troops happy, see pages 35–36. Some beneficials may establish themselves permanently in the garden, while others will do their duty, then move on or die out when conditions no longer suit them.

Here and on the following pages, you'll find descriptions of some of the natural enemies available to home gardeners.

Aphid Midges

In its larval stage, this tiny black fly (*Aphidoletes aphidimyza*) feeds on many types of aphids, especially those infesting rosebushes, apple trees, and shade trees. The adults aren't predators—they eat honeydew, the undigested plant sap excreted by aphids.

Aphid midges are sold as pupae, which you distribute on the soil or on plants. The minuscule adults emerge and mate; then the females lay orange eggs near aphid colonies. Legless orange larvae, about ⅛ inch long, hatch out within a few days and begin feeding, sucking the aphids' body fluids and leaving the shriveled corpses behind on your plants. The larvae continue to feed for a week or two, going to work when the sun sets. They then pupate in the soil for a couple of weeks, after which the cycle starts all over again. You will need 100 to 300 pupae for about 1,000 square feet.

Convergent Lady Beetles

These insects (the scientific name is *Hippodamia convergens*) are the most popular natural enemies sold, though their value as released beneficials is doubtful: they have a natural tendency to disperse when freed. To increase the chances of hanging on to your helpers, you might try watering the garden, providing a commercial insect food, and scheduling the release at dusk (the beetles don't fly at night). Some suppliers recommend spraying the beetles with a soft-drink solution to "glue" their wings and prevent them from flying for about a week.

Aphids are the food of choice for both the familiar orange- and black-spotted adult beetles and the alligatorlike larvae, which sport numerous raised black spots on a blue-gray and orange background. (See photos on page 37.) If the aphid population is too low, larvae and adult beetles may also eat scales and other soft-bodied insects, mites, and insect eggs. Adults feed on nectar, pollen, and honeydew as well.

Most suppliers recommend releasing from 1 cup to 1 quart of beetles in home gardens, freeing the insects gradually over several evenings. Rough handling can spur lady beetles to flight, so place them gently at plant bases.

Decollate Snails

Rumina decollata glides out at night to feed on young brown garden snails and snail eggs. Unlike its round-shelled prey, the decollate snail has a conical shell with a squared-off tip. It reaches a length of about 2½ inches.

Besides dining on brown snails, decollate snails eat rotting vegetation, seedlings, and—if they're hungry—tender plant parts in contact with the soil. For this reason, it's best not to release the snails in vegetable or flower beds. Until the decollates establish control, you may still need to handpick brown snails; don't use snail bait, since decollate snails are just as susceptible to it as their pesky brown cousins.

Releases of one or two decollate snails per square

The larger decollate snail preys on its smaller cousin, the pesky brown garden snail.

yard usually suffice, although it may take several years before brown garden snails are entirely displaced.

If you live in California, be aware that decollate snails are legal only in the nine southern counties; elsewhere in the state, their use is banned due to concerns about the potential impact on endangered mollusks.

Green Lacewings

Ethereal, golden-eyed adult lacewings (see page 37), their clear wings extending beyond their light green bodies, nibble only on nectar, pollen, and honeydew. The predators—and fierce ones at that—are the grayish brown larvae (see page 4). Resembling ½-inch-long alligators with pincerlike mouthparts, they suck fluids from insect eggs and soft-bodied prey such as aphids, thrips, leafhopper nymphs, small caterpillars, immature whiteflies, soft scales, mealybugs, and spider mites. When food supplies run short, they turn cannibalistic, preying on other lacewing larvae and eggs.

Lacewing larvae feed voraciously for about 3 weeks, then pupate in round white cocoons that are usually attached to leaf undersides. The adults emerge in about 5 days, then feed for a few days before laying eggs. The cycle probably won't continue in cool climates.

The most common lacewing species sold, *Chrysoperla carnea* (formerly *Chrysopa carnea*), is good for general garden use. *C. rufilabris*, an eastern species, is often released in orchards and into shade trees. Some suppliers sell a mixture of the two species.

A few companies ship lacewing larvae, but most send eggs; killed caterpillar or beetle eggs serve as food for any larvae that hatch in transit. To minimize cannibalism further, suppliers use rice hulls, vermiculite, or other material to separate the eggs. Distribute eggs or the newly hatched larvae on plants, allowing one to five per square foot of garden space.

Mealybug Destroyers

Both the larvae and adults of this Australian lady beetle (*Cryptolaemus montrouzieri*) devour mealybugs. If they can't find enough of their favorite food, they may also eat aphids and immature scale insects. The ⅛-inch-long adults have black bodies and orange heads (see below); the pale, waxy-coated larvae are hard to distinguish from mealybugs (see page 15).

Each adult female lays hundreds of eggs, depositing them singly in mealybug egg masses. When the beetle larvae hatch, they feed on immature mealybugs.

Mealybug destroyers need warm temperatures and high humidity, so they're better suited to greenhouses than gardens, although they do thrive outdoors in some areas. Short days and temperatures below 68°F slow their rate of reproduction. They don't survive cold winters.

The insects are shipped as adults. Once they arrive, release them as soon as possible; don't refrigerate them. Suppliers recommend distributing one or two

Mealybug destroyers—whether larva or adult (the black and orange beetle shown here)—feed on mealybugs.

Dealing with Pests & Diseases **53**

This package of parasitic nematodes (above) comes packed in a gelatin-coated screen. The nematodes are mixed with water according to the manufacturer's directions and applied with a watering can (below).

beetles per square foot of planted area or two to five per infested plant.

Parasitic Nematodes

While some nematodes are extremely destructive (see page 85), others are helpful. Barely large enough to be seen with the naked eye, these roundworms seek out and enter more than 400 kinds of soil-dwelling and boring pests. They also attack insects that pupate in the soil. However, they're harmless to earthworms.

As they feed, nematodes release bacteria that slowly kill the host. Most pests die in a few days, though they may linger up to a month in cool conditions. Nematodes complete their life cycle inside the host's dead body; then the offspring leave to seek new hosts.

The nematode species most commonly sold are *Sc* and *Hb*—short for *Steinernema carpocapsae* (formerly *Neoaplectana carpocapsae*) and *Heterorhabditis bacteriophora* (formerly *H. heliothidis*). Various strains of each species are available. *Sc* strains tend to stay near the surface, successfully controlling pests like cutworms and armyworms, while *Hb* strains usually move a little deeper into the soil (to about 6 inches) and are effective against certain beetle grubs. Different strains vary in their ability to kill specific pests. To decide which species is suitable for your needs and to determine application rates, consult the supplier. Don't apply more than one kind of nematode in the same soil, since different types can't seem to coexist successfully.

Sealed boxes of nematodes will keep for months in the refrigerator. Packages typically contain from 10 million to 3 billion nematodes, held in a gel, clay, or sponge medium or mixed with moistened peat and vermiculite.

Adequate moisture and proper placement are the keys to success in using nematodes. Mix the nematodes with water, then apply directly to the soil with a watering can or sprayer; or use a garden syringe or oilcan to shoot the mixture into tunnels bored in stems and trunks. Apply nematodes after sunset (so they won't be exposed to sunlight) or in the daytime during a light rain. For best results, keep the treated areas moist. It's also important to apply nematodes regularly; though they can persist in moist soil for long periods, even in the absence of hosts, they're most effective in the 2 to 3 weeks following application. Avoid using a nematicide against pest nematodes, since it's lethal to helpful species as well.

Parasitic Wasps

Generally too small to be noticeable, these miniwasps don't bother people or pets. Several species are already available to home gardeners, and many more are under study and likely to be marketed in the future. The wasps are shipped as mature pupae in parasitized hosts (usually glued to pieces of cardboard) and will emerge as adults soon after you receive them. When distributed among plants, they'll seek out hosts in which to lay their eggs.

The whitefly parasite (*Encarsia formosa*) is a minute wasp that lays eggs in the scalelike third and fourth nymphal stages of greenhouse and sweetpotato whiteflies. The infested scales blacken as the wasps develop inside. Adult wasps also feed directly on whitefly nymphs. Although most effective in a confined space such as a greenhouse, these wasps can be released in warm, sunny gardens sheltered from winds. When temperatures fall below 62°F, however, they'll stop flying and thus fail to find hosts. Suppliers recommend releasing the wasps weekly for 8 to 10 weeks, at a rate of two to four wasps per square foot of garden space or one to five wasps per infested plant.

Trichogramma wasps are so tiny that four or five of them can perch on the head of a pin. Of the many *Trichogramma* species, some are generalists, while others prefer only one or a few related hosts. Three species are commonly available for purchase: *T. pretiosum*, for general garden use; *T. platneri*, for tree crops on the West Coast; and *T. minutum*, for tree crops elsewhere. *T. pretiosum* infests the eggs of more than 200 species of butterflies and moths, including cabbage loopers, codling moths, corn earworms, and hornworms. As the wasps develop, they kill the embryos inside the host eggs, thus preventing pest damage before it can begin. A parasitized egg will turn dark—and when it hatches, an adult wasp will emerge instead of a caterpillar.

Release trichogramma wasps at a rate of about one per square foot, at the same time the target pest begins laying its eggs (consult your supplier or Cooperative Extension agent to determine the right release date). The adults will emerge within a few days and remain active for about 9 days. Because a shipment can be stored for several weeks, releases can be staggered.

The bean beetle parasite, also known as the pedio wasp (from its scientific name, *Pediobius foveolatus*), lays eggs in Mexican bean beetle larvae. Time the release to coincide with the pest's larval stage. One shipment is enough for a 400-square-foot planting.

Praying Mantises

Egg cases of praying mantises are widely sold, but these insects aren't much help in pest control. They're indiscriminate predators, dining on beneficial insects, pests, and even on each other. There's only one generation a

*This parasitic wasp (*Trichogramma pretiosum*) is laying an egg in a corn earworm egg.*

year, and most of the nymphs die young. Furthermore, the insects reproduce more successfully when they arrive naturally rather than when they're introduced into a garden. *Tenodera aridifolia sinensis* is the species most commonly sold. (See photo on page 38.)

Predatory Mites

Home gardeners can buy various species of predatory mites. Typically transparent to salmon in color, these helpful creatures move about quickly, gobbling pest mites; the nymphs feed on pest nymphs and eggs, while the adults consume all pest stages. Once the food supply runs out, the predators starve to death. Suppliers usually send *Phytoseiulus persimilis*, a species especially useful in herbaceous plantings.

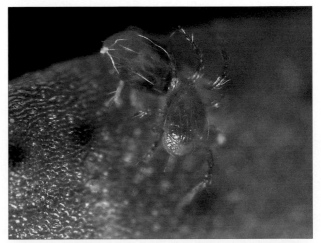

This greatly enlarged view shows an agile, orange-hued predatory mite attacking a slower-moving spider mite.

Two predatory mite species, *Amblyseius cucumeris* and *A. barkeri*, feed mainly on thrips, specifically on western flower thrips and onion thrips.

You'll need about two predatory mites per infested leaf. The mites are generally shipped as adults and should be released as soon as they arrive.

Spined Soldier Bugs

This true bug (*Podisus maculiventris*) is sold primarily as a control for the Mexican bean beetle, but it also feeds on just about any insect larvae or eggs exposed on a leaf surface. It attacks more than 100 kinds of garden pests, including gypsy moth caterpillars, Colorado potato beetles, corn earworms, armyworms, and cabbage loopers. Soldier bug nymphs and adults are equipped with long, daggerlike mouthparts, which they keep folded beneath their bodies when not feeding. To feed, the bugs stab prey with their mouthparts, then suck out the victim's body fluids. If sufficient prey is available, soldier bugs do a good job of controlling pests for several weeks after release.

Spined soldier bugs are shipped as eggs, which may hatch en route into round, wingless nymphs in several shades of orange with black markings. The nymphs develop into ½-inch-long, grayish brown adults bearing the characteristic shield shape and triangular marking of true bugs.

Suppliers advise distributing 100 eggs along every 20 feet of row. If spined soldier bugs aren't common in your area (they're most prevalent in the East), putting out pheromone lures may increase the odds that they'll stay. (For more information about pheromones, see page 47.)

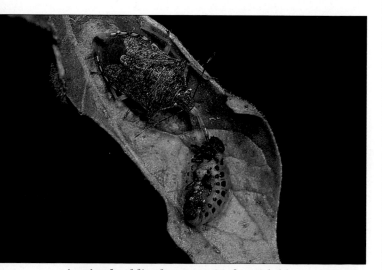

A spined soldier bug uses its formidable mouthpart to spear a Colorado potato beetle larva.

Pathogens

Often referred to as microbial pesticides, these biological controls consist of living microorganisms or the toxins they produce. Selectivity and safety are their greatest strengths: they kill targeted pests without harming other creatures or treated plants, and can even be applied right up until harvest.

You may have to apply some pathogens repeatedly, but others reproduce on their own to provide long-term control. Because sunlight and heat decrease the effectiveness of some microbials, proper application according to package directions is important. The shelf life of certain products may also be limited; again, check the label.

In addition to the widely merchandised *Bt* (the full name is *Bacillus thuringiensis*) and milky spore disease, many garden suppliers sell *Nosema locustae*, a grasshopper pathogen. Viruses that attack certain pests are marketed as well, but these aren't yet generally available to home gardeners. You can also make your own microbial by liquefying diseased pests and spraying them on infested plants.

Bt

Widely used since the 1960s, *Bt* (*Bacillus thuringiensis*) is a stomach poison: it paralyzes and destroys the stomach cells of the insects that consume it. Though the poisoned pest may not die for several days, it soon stops feeding, bringing plant damage to a halt. Because *Bt* must be ingested to work, it's often mixed with a commercial appetite stimulant to encourage pests to eat more.

Unlike some other pathogens, *Bt* usually doesn't reproduce and infect succeeding pest generations; it must be reapplied to control future outbreaks. Most products are effective for one to a few days, although new encapsulated versions last longer.

Bt is available in various formulations: spray, dust, and granules (some granular forms are baited to entice the pest). You'll get better results if you don't mix the spray concentrate with alkaline water (pH 8 or higher), since alkalinity reduces its effectiveness. (To acidify water, stir in a couple of teaspoons of white vinegar per gallon before adding the *Bt*.)

For maximum control, cover all plant surfaces thoroughly with spray. Schedule applications in late afternoon or evening or on an overcast day, so the *Bt* has a chance to work before exposure to sunlight inactivates it.

Until recently, commercial *Bt* products were successful only against leaf-feeding caterpillars, but now you can also buy strains that control certain other pests. New strains are constantly being discovered and will doubtless be available for sale in the future.

Bt caterpillar toxin is sold under many trade names. Use it sparingly—and if you want butterflies in your garden, apply it only to plants hosting pest caterpillars. The pests controlled include bagworms, cabbage loopers, cankerworms, fall webworms, gypsy moth caterpillars, hornworms, imported cabbageworms, and tent caterpillars. One of the available formulations is particularly effective against armyworms, although it also works against all the usual caterpillars. Since *Bt* works best on young larvae, apply the toxin as soon as you notice that eggs have begun to hatch.

Caterpillar toxin is ineffective against soil-dwelling caterpillars and those that bore into fruit before eating enough of the treated surface to be poisoned. However, you may be able to kill stem and trunk borers by injecting toxin into the holes.

Bt san diego, a relatively new strain, is toxic to certain beetles. Now available in encapsulated form, it's deadly to young Colorado potato beetle larvae and to elm leaf beetle larvae and adults. Strains lethal to other beetles are expected to be available in the future.

Bt israelensis is fatal to mosquito, black fly, and fungus gnat larvae. It's most often used in community-wide mosquito abatement programs; in a home garden, an easier and more permanent solution is to eliminate standing water. If mosquitoes are a problem in your ornamental pool, adding a few goldfish or mosquito fish should keep the insects under control.

Milky Spore Disease

This pathogen contains the bacteria *Bacillus popilliae* and *B. lentimorbus.* Applied to lawns and watered in, it infects Japanese beetle larvae in the soil. It also controls, albeit less effectively, certain other beetle grubs.

Once ingested, this microbial slowly liquefies the Japanese beetle grub's internal organs and turns them milky white—hence the name "milky spore disease." When the insects die, sometimes after weeks, the bacteria are released into the soil, where they infect other grubs. It usually takes about 3 years to achieve control, but after that, the microbial can remain active for 15 to 20 years.

Milky spore disease can be used whenever the ground isn't frozen, but it's best applied in spring and fall, when the grub population is high. Results are best if entire neighborhoods cooperate; being a lone user won't do you much good, since adult beetles can fly in from adjacent properties.

Pest caterpillars, such as this hornworm, stop feeding soon after ingesting Bt-*sprayed foliage, although they may take several days to die.*

Nosema locustae

A naturally occurring disease that afflicts 58 species of grasshoppers, this pathogen has been packaged and is sold under various trade names. It's used primarily on rangelands; it's not too practical for the average home garden, since it works very slowly and won't prevent immediate damage to crops.

Nosema locustae works best when ingested by very young nymphs. Though only about half the grasshoppers die (usually within 3 to 6 weeks of infection), the disease slowly brings the pest under control by reducing feeding and reproduction among the survivors. The illness also spreads when healthy grasshoppers eat their ailing compatriots.

Liquefied Pest Sprays

To make your own pathogen, grind up diseased pests with a little water in a blender. Since most viruses and other disease organisms are very host-specific, collect just one type of pest (cabbage loopers and slugs are among those commonly chosen) and use them against other members of the species. After grinding up the pests, dilute the liquid with more water; then spray the infested area.

Collecting only diseased pests isn't always easy, since some infected individuals appear completely normal. To increase your chances of getting sick pests, just gather as many as possible—regardless of how ill or healthy they look. The spray may do a good job even if none of the pulverized pests was diseased: according to one theory, the smell of liquefied bugs attracts natural enemies.

CHEMICAL CONTROLS

hemical dusts, sprays, and granules are the pest-control agents of last resort, to be applied judiciously when less toxic methods have failed and the level of damage is expected to be intolerable. The materials included in this category range from the benign to the downright dangerous. Thoughtful selection and use will allow you to control the pest without harm to yourself, those around you, or the environment; see the section on pesticide safety on pages 65–67.

Prior to World War II, home gardeners solved most of their pest problems with soaps, oils, botanical (plant-derived) pesticides, and inorganic (non-carbon-containing) materials like copper and sulfur. Then came DDT, a powerful synthetic insecticide that ushered in a new, "take-no-prisoners" attitude toward pest control: the only acceptable solution to pest problems was total eradication. Interest in other approaches waned almost overnight.

The first laboratory-produced insecticides were organochlorines, most of which have since been banned due to the long-term hazards they present to people, wildlife, and the environment. Other types of synthetic insecticides—organophosphates, carbamates, and pyrethroids—are still widely available, although some have been prohibited or restricted. Many of the fungicides turned out to be possible carcinogens and were outlawed for that reason.

The removal of so many synthetics from the market has sparked a return to natural remedies, even though these products are not without risk. Today, the conscientious home gardener can choose from a variety of natural and synthetic controls that work effectively if used wisely.

Each pesticide discussed in this chapter is referred to by its common name or active ingredient. Brand names and promotional copy alone may not tell you just what a particular commercial product is; to find out whether it contains pyrethrum, carbaryl, or some other chemical, you may have to consult the list of active ingredients on the label.

The chemicals reviewed here serve specific purposes. They're not interchangeable. An insecticide won't cure powdery mildew, a fungicide won't kill snails, and a soap spray won't penetrate hard-shelled beetles. A contact poison won't work against borers once they've tunneled inside a plant, and a stomach poison sprayed on leaves won't kill a sucking pest that doesn't eat the outer plant surface. Before applying a material, make sure it's appropriate for the pest you want to control. Some creatures are susceptible to different materials at different times, so be flexible.

In commercial growing fields, it's important to vary the chemicals used (especially the synthetics), since repeated exposure to the same pesticide can breed resistance in a pest species: in each generation, a few tolerant individuals survive to reproduce, eventually giving rise to a more tolerant population. Resistance is less likely to arise in a typical home garden, given the fact that insects constantly migrate in and out. Still, alternating controls is a good practice.

Any product sold to control pests must be registered both at the federal level by the Environmental Protection Agency (EPA) and at the state level by each state's Department of Agriculture or similar agency. Some products used to kill pests—dishwashing detergent, for example—may qualify as pesticides in a gardener's eyes, but legally they aren't pesticides if they make no claim to control pests.

Certain products may not be available in your area. If the registration process in a given state is lengthy and expensive, or if the targeted pest isn't a problem there, a manufacturer may decide not to register a product in that state. Suppliers in other states may offer the product by mail, but strictly speaking it should only be shipped to states in which it's legally registered. You can always write to the manufacturers or distributors; if there's enough demand, they'll probably register the product in your state.

Mineral-based Products

Mineral-based products for pest control consist entirely or primarily of mined elements. All are inorganic chemicals—but paradoxically, most are accepted by organic gardeners.

Sulfur

Elemental sulfur—the ground mineral—is one of the oldest weapons against plant diseases, mites, and some insects. Modern Greek grape growers dust their vines with sulfur, just as their ancestors did thousands of years ago. Ancient Egyptians and Sumerians also used sulfur on their crops.

Sulfur protects plants from fungus infections, but it doesn't cure infected tissue. It's useful against mildews, rusts, brown rot of stone fruit, gray mold, black spot, and certain other diseases. Though sometimes used against aphids, scales, and thrips, it's far more effective against mites (including predatory types). Because it's less toxic to insects than are many other pesticides, it presents less danger to beneficials than other products do. To be on the safe side, however, don't release beneficial insects until a week after applying sulfur.

Modern sulfur formulations include dusting sulfur, wettable powder, flowable sulfur, and lime sulfur (a caustic mixture sprayed on dormant fruit trees). You'll also find combination formulas of sulfur and other pesticides, although the sulfur usually has a greater tendency to burn plants when mixed with other materials. Sulfur used in combination with oil is especially likely to cause burning, so wait to apply sulfur until a month after using an oil spray. Also avoid applying sulfur when temperatures exceed 90°F.

Don't use sulfur on sensitive crops, including melons, cucumbers, squash, pumpkins, apricots, and raspberries. If you intend to can your produce, don't apply sulfur immediately before harvest: sulfur dioxide can form, causing the sealed jars to burst.

Sulfur has a low toxicity to humans, although some people are allergic to it. Repeated applications will acidify the soil, a desirable side effect if your soil is too alkaline.

Copper Compounds

Various copper compounds are used to control many fungal and bacterial diseases. Basic copper sulfate, the most common, is available as dust, wettable powder, and liquid concentrate.

Bordeaux mixture, named for the grape-growing region of France where it was developed over a century ago, is a combination of copper sulfate and lime. Growers who painted the nasty-looking blue mixture on roadside grapevines to discourage pilferers discovered that it also protected the plants from mildew. Sold as a wettable powder, Bordeaux mixture is best when made fresh for each use.

Bordeaux mixture was applied to these grapevines to protect them from mildew.

Some plants are sensitive to copper; if you have questions about the particular plants you intend to treat, consult your Cooperative Extension agent or an experienced nurseryman. Also be aware that copper can cause damage if applied during hot weather or at too high a rate on young plants.

Copper has a low toxicity to humans (though it can cause eye and skin irritation), but it harms fish if it gets into waterways. Don't overuse it, since it can build up to toxic levels in the soil.

Baking Soda

Sprayed onto plants, a solution containing ordinary baking soda prevents certain fungus diseases, including black spot, powdery mildew, and anthracnose. Scientists aren't sure just how this control works—but it's so effective against a wide variety of fungi that efforts are under way to register baking soda legally as a fungicide.

To make the solution, mix 1 tablespoon of baking soda (sodium or potassium bicarbonate) and 2½ tablespoons of ultra-fine oil spray with a gallon of water. The oil is slightly fungicidal and acts as a spreader-sticker, helping the baking soda coat the leaf and cling to the surface longer. The solution is a contact protectant: it stops fungi from attacking plant tissue but doesn't cure tissue that's already infected. Apply as soon as symptoms appear; repeat every 2 weeks or so. Be sure to coat all plant surfaces.

Diatomaceous Earth

Diatomaceous earth consists of the skeletal remains of diatoms—single-celled marine algae. It kills soft-bodied pests by lacerating their bodies and causing them to dehydrate. Diatomaceous earth is most often used as a border around planting beds to keep out slugs and snails. It may also be sprinkled around the base of individual plants, dusted on foliage, or mixed with water to make a foliage spray. Applied on plants, it works against pests such as aphids and mites.

If you plan to dust your garden with diatomaceous earth, do so during hot, dry weather. The material is most effective then—and you'll be spared the bother of having to reapply it after each rainfall.

Diatomaceous earth doesn't hurt mammals, birds, or earthworms, although it can harm beneficial insects; to protect bees, avoid coating flowers. Wear protective gear when applying diatomaceous earth, since it can irritate eyes and lungs. Be sure you buy natural or agricultural diatomaceous earth—not the pool-grade product, which has smoother-edged particles and lacks lacerating power.

Chitin

A material found in crustacean shells, chitin represents a home gardening breakthrough in pest nematode control. Various chitin-based products, made from ground-up shells and usually combined with nitrogen, are actually fertilizers that happen to kill nematodes as well. Although quite pricey, they're a safe alternative to the highly toxic soil fumigant sold as a nematicide. If you're allergic to shellfish, however, be careful with chitin-containing products.

Chitin products work by promoting the growth of soil-dwelling microorganisms that feed on chitin—and also on nematode eggshells, which contain chitin.

Since most pest nematodes attack plant roots, chitin works best when dug deep into the soil, as far down as roots will grow. It can also be watered into lawns and perennial beds. The product should be applied annually; the initial application is usually 4½ pounds per 100 square feet, but you may need to double or triple the dose in heavily infested or very acidic soils. In subsequent years, application rates can be lower. After treating soil with chitin, wait a couple of weeks before planting, since high levels of quick-release nitrogen can burn plants. In the interval, keep the area well watered.

Soaps

Insecticidal soaps are mild poisons made from the salts of fatty acids found in plants and animals. They kill by penetrating the cell membranes of soft-bodied pests such as aphids, whiteflies, thrips, leafhopper nymphs, scale insects (crawler stage only), and spider mites. Soaps aren't successful against pests with tough body coverings, nor are they effective on fast-moving insects that can evade the spray. Most adult beneficial insects can endure or outfly a soap spray, but their soft-bodied, flightless larvae may be injured or killed.

For additional killing power, commercial sprays may supplement the soap with other ingredients, such as pyrethrins or citrus oils. Soap-sulfur sprays are also sold; these serve as an insecticide, miticide, and fungicide, all in one. Commercial products are formulated to be toxic to pests with minimum harm to plants, but if you prefer to make your own spray, just combine 2 tablespoons of liquid dishwashing detergent with a gallon of water.

Since soap sprays kill on contact, you must score a direct hit on the pest. It's important to coat all plant surfaces, especially leaf undersides and crannies where your quarry may be hiding. Also remember that soap works only when wet; to prolong its effect, spray early

Soap sprays are among the least toxic chemical controls available to home gardeners. You can buy a commercial product or make your own solution from dishwashing detergent.

in the morning or late in the day, when the spray will dry slowly. To further delay drying, add ¼ teaspoon of any vegetable oil to each quart of soap spray.

When mixing your own spray or diluting a commercial concentrate, use warm soft water to ensure that the soap mixes as uniformly as possible. To find out if your water is soft, add a few drops of the soap to a quart of water, then shake. If the solution is clear and sudsy, the water's soft; if it's whitish with few suds, the water's hard. If necessary, add a water softener according to label directions; or collect rainwater for use with soap sprays.

Soap leaves no residue and is safe to use on edibles. However, some plants are soap-sensitive, so test any spray—purchased or homemade—on a small section of the plant if you're at all unsure of the possible effects. Wait a couple of days, then inspect for damage. Also check commercial product labels for a list of plants that may be harmed.

Oils

Most oils used for horticultural purposes are petroleum based, although some are made from vegetable and fish oils. These products kill insects and mites—both harmful and beneficial types—either by smothering them (and their eggs) or by interfering with membrane functions.

Oils have been a standard weapon in the home gardener's arsenal for more than a century. The original formulations could only be used on deciduous plants during the dormant season; at other times of year, they'd burn plants or even kill them.

In the last 15 years or so, new refining techniques have made it possible to produce lighter oils, some of which can be used all year. Called summer, superior, or verdant oils, these products are applied during the growing season to control many pests, including lacebugs, aphids, scale insects, whiteflies, mealybugs, and spider mites. Some summer oils actually help prevent powdery mildew, rust, and gray mold on roses, lilacs, hydrangeas, phlox, crape myrtle, zinnias, and other ornamentals.

Even summer oils can burn sensitive leaves, so test the spray on a small area before coating the entire plant. It's best to use these oils when the outdoor temperature is between 40° and 90°F.

Just how a summer oil behaves depends on its distillation temperature. The lower the temperature, the faster the oil evaporates and the less likely it is to cause plant damage. For best results, use oils distilled between 412° and 414°F. (By contrast, most dormant oils are distilled between 438° and 476°F, multipurpose oils at 435°F.) The distillation temperature isn't usually listed on the label, so you may have to ask the supplier or manufacturer. Keep in mind that many dormant oils have simply been relabeled for use as summer oils, but with lower application rates during the growing season. They're not the same as true summer oils.

To make your own inexpensive, nontoxic oil spray for use on growing plants, use the following USDA formula; it has been found to be as effective as two commercial products (an insecticidal soap and a soybean oil sold for pest control) in killing aphids, whiteflies, spider mites, and other pests. Combine 1 cup of oil (peanut, safflower, corn, soybean, or sunflower) with 1 tablespoon of liquid dishwashing detergent. To make the spray, use 1½ teaspoons of the oil-detergent mixture for each cup of water. Coat all plant surfaces thoroughly; if necessary, repeat the application in 7 to 10 days. This spray may burn cauliflower, red cabbage, and squash leaves.

Botanical Pesticides

These natural, plant-derived pesticides are generally less environmentally harmful than synthetic pesticides because they don't persist as long: when exposed to sunlight, they break down into harmless substances within hours or days. A botanical pesticide's fleeting life makes the product more difficult to use effectively, however. Your timing and aim must be precise, and you may have to apply the material more frequently to get the results you'd achieve with a more persistent pesticide.

Botanical pesticides aren't necessarily "safe" just because they're natural. Some of these products are very toxic and can harm anyone using them carelessly. In fact, some have a higher toxicity rating than many synthetics; see the chart on page 66. Certain botanicals are also toxic to pets, wildlife, and beneficial garden creatures.

A botanical pesticide usually consists of crude plant material ground into a dust or wettable powder; it may also be made from plant extracts or resins

Pyrethrum products are derived from the colorful flowers of the pyrethrum daisy, a type of chrysanthemum.

formulated into a liquid concentrate or impregnated onto a dust or wettable powder.

Use of crude botanical pesticides dates back centuries. Pyrethrum was a traditional pest control in Persia, rotenone in South America, and neem in India. Botanicals were widely used in the United States from the late 1880s until the 1940s, when they were nearly abandoned in the excitement over the newly developed synthetics. Today, however, home gardeners and small organic farmers are flocking back to botanicals.

Botanicals are sometimes combined with synthetics, so read the label carefully if you want the natural product alone. Also be aware that botanicals often contain synergists, laboratory-produced compounds that block an insect's ability to detoxify the pesticide. Synergists are low in toxicity and have only brief residual activity, but they're not natural and organic gardeners may not want to use them.

The major botanical pesticides available today are discussed below.

Pyrethrum Products

Pyrethrum products are probably the most widely used botanical pesticides. Pyrethrum is a dust made from the whole flower of the pyrethrum daisy (*Tanacetum cinerariifolium*, formerly called *Chrysanthemum cinerariifolium*); pyrethrins are the naturally occurring toxins extracted from the seeds and used in making various insecticidal products. Both are natural materials. Pyrethroids, on the other hand, are synthetics that are more toxic and persist longer in the environment than pyrethrum or pyrethrins.

Pyrethrum products kill a broad range of pests by paralyzing them on contact; they're especially noted for their rapid knockdown of flying insects. Susceptible pests include aphids, beetles, caterpillars, leafhoppers, mealybugs, thrips, and whiteflies. Beneficial insects may also be killed, but this drawback is mitigated by the fact that the toxin breaks down within a few hours of exposure to sunlight. Applying the pesticide in the evening lessens the risk to bees.

A pyrethrum product is most effective when dusted or sprayed directly on the adult insect. Nonetheless, the pest may only be stunned if it gets too little toxin or can break down what it does receive. For this reason, most pyrethrum products contain the synergist PBO (piperonyl butoxide), which prevents insects from metabolizing the poison. Some formulations contain not PBO but soap, which supposedly intensifies the pyrethrum's effect.

You can also get a microencapsulated version of pyrethrum in a spray can; it's effective for up to a week outdoors. Remember, though, that this longer-lasting poison presents a greater danger to bees and other beneficial insects.

Many pyrethrum products also contain rotenone, ryania, or both—the fast-acting pyrethrum for the knockdown and the slower-acting rotenone or ryania for the kill. Pyrethrums are also sold in combination formulas containing synthetic pesticides or fungicides like copper.

Pyrethrum has a low toxicity to humans, and the more concentrated pyrethrins are only slightly more toxic. These products are more harmful when inhaled than when ingested, and the dusts may also cause skin irritation and/or allergic reactions, so include a mask and protective clothing in your application equipment. If you're sensitive to chrysanthemums, avoid pyrethrum products.

Pyrethrum products will poison fish and other aquatic life, so don't use them around waterways.

Rotenone

A broad-spectrum contact and stomach poison, rotenone is derived from the roots of several tropical legumes, including derris and cubé. It interferes with cellular respiration, causing death within a few hours to a few days after exposure. Once applied, the material remains active for 3 to 7 days.

Rotenone kills many types of pests—most notably leaf-feeding beetles and certain caterpillars, but also harlequin bugs, squash bugs, thrips, scale insects, and leafhoppers. Before the development of *Bt san diego* (see page 57), rotenone was widely used in the Northeast to control the Colorado potato beetle, which had become resistant to other chemicals.

Among the most potent of the botanical pesticides, rotenone is more toxic to vertebrates than either malathion or carbaryl, two of the most widely used synthetics. However, humans and most other mammals (except pigs) are able to detoxify ingested rotenone. If inhaled, the material is more toxic, so always wear a mask when applying rotenone dust. Direct contact can also irritate the skin and mucous membranes.

Rotenone is deadly to fish and other aquatic life; native tribes of South America traditionally used it to kill fish, which they then safely consumed. Rotenone is toxic to some beneficial insects, but it doesn't harm bees.

You can buy this pesticide as a dust or a wettable powder for spraying. It's also available in combination formulas with pyrethrum, ryania, copper, and sulfur. (When combined with pyrethrum, it's toxic to bees.)

Bees are the major pollinators in gardens—so use products that spare them or that dissipate quickly. If you use a chemical that's harmful to bees, apply it in the evening, when the insects are in their hives.

Ryania

This slow-acting stomach poison consists of ground-up woody stems of the tropical ryania shrub. After consuming the poison, pests don't drop dead immediately—they just get too sick to eat the treated plants. Since bees and other beneficial insects don't feed on plants, they're safe from ryania's effects.

The most selective of the botanical pesticides, ryania is registered for use on citrus, corn, walnuts, apples, and pears—as a control for citrus thrips, European corn borers, and codling moths.

Ryania remains active for up to 2 weeks after application. Since it washes off plants, it's most effective when the weather is warm and dry.

This pesticide is usually sold as a dust. Although the concentrated material is highly toxic to mammals, most formulations are fairly low in toxicity.

There are very few sources for ryania in pure form; it's usually sold in combination formulas with rotenone and pyrethrum, to control a broader range of pests.

Sabadilla

A broad-spectrum pesticide made from the seeds of a South American lily, sabadilla is effective against certain hard-to-control true bugs, including squash bugs, harlequin bugs, tarnished plant bugs, and chinch bugs. It also kills other insects, such as leafhoppers and cucumber beetles.

Primarily a contact poison, sabadilla also has some effect as a stomach poison. It works best when applied directly on the adult insect. Certain kinds of pests will be killed outright, but others may be paralyzed for days before dying.

When sabadilla is exposed to sunlight and air, its toxic components break down quickly—in about 2 days. You may have to apply the pesticide weekly until the targeted pests are under control.

Sabadilla dust made from ground seeds is one of the least toxic botanicals, but purified formulations (purified veratrine alkaloids extracted from the seeds) are just as toxic as the deadliest of synthetics. Sabadilla is poisonous to bees, so apply it in the evening when bees are less active.

When applying dust or spraying wettable powder, wear protective gear; the material is very irritating to the skin and mucous membranes.

Neem

Derived from the tropical neem tree (*Azadirachta indica*), this botanical pesticide is brand new on the

American market, but it has a long history in India and other parts of the world. Every part of the source tree repels pests, and neem is correspondingly effective against a wide range of garden troublemakers—but it's virtually nontoxic to humans and most other mammals. In India and other Asian countries, neem is used in medicines, cosmetics, and household products.

Neem is a successful weapon against beetles, caterpillars, grasshoppers, crickets, and many sucking insects (with the exception of scale insects and psyllids). The active ingredients are azadirachtins, which are growth regulators, feeding deterrents, and repellents. As a growth regulator, neem interferes with the ability of the larval and pupal stages of some insects to molt and reach reproductive maturity. Other insects won't feed on plants treated with neem (nerves associated with feeding are disrupted); still others will avoid treated plants entirely.

Neem is currently registered for use only on ornamentals, but manufacturers are seeking approval for its use on edibles as well. Leaf sprays are available now, and other formulations are expected to follow. Apply neem approximately every 7 days for prevention, every 3 or 4 days if pests are present. The material breaks down in about a week.

The word "neem" may not appear on labels at all, so look for azadirachtin in the list of active ingredients.

Nicotine Sulfate

Extracted from tobacco plants, nicotine is so poisonous that it's not available to home gardeners in its pure form. Suppliers sell only nicotine sulfate, a formulation that's slightly less toxic but still potentially fatal if inhaled, ingested, or absorbed through the skin. Nicotine sulfate kills soft-bodied sucking pests (such as aphids, thrips, and spider mites) on contact. It's also deadly to many beneficials.

Because less toxic alternatives are available, nicotine use has been sharply curtailed. In fact, it is now restricted (limited to professional application or to use on particular pests or plants) in many states.

Synthetic Pesticides

For several decades after World War II, synthetics played the starring role in pest control—to the virtual exclusion of all other methods. Eventually, however, these "wonder" chemicals were found to have serious drawbacks. Some left undesirable residues in the environment; some proved to be carcinogens. Many were pulled from the shelves.

Still, many synthetic products are relatively safe and effective when used properly—and in concert with other control methods. Gardeners who rely solely on synthetic chemicals risk killing helpful creatures—and thus inadvertently fostering populations of pests that the beneficials previously kept in check. Repeated use of the same type of chemical can also encourage the development of pest strains resistant not only to an individual product, but to an entire class of chemicals.

On the positive side, most synthetics can be applied less often than other types of pesticides, since they usually remain effective longer.

Insecticides

The following are the most commonly used synthetic insecticides. All are broad-spectrum products—that is, they kill a wide range of insects.

Malathion. Sold as a liquid concentrate, this organophosphate is deadly on contact, killing aphids, mealybugs, scale insects, leafhoppers, whiteflies, cucumber beetles, and many other insects that feed on exposed plant surfaces. Because it's effective for just 1 to 3 days after application, malathion is less damaging to beneficial insects than are many other synthetics.

Probably the safest of the widely used synthetics, malathion is about as toxic to humans as ryania and pyrethrum; it's somewhat less toxic than rotenone. Malathion is more toxic when absorbed through the skin than when ingested, so cover up well when applying it. Also be careful not to store it in a very hot shed or garage, since its toxicity increases at high temperatures.

Diazinon. Another contact killer, this organophosphate is available in liquid concentrate, dust, or granular form. It's often recommended for controlling chewing and sucking pests in vegetable gardens and orchards. Diazinon lasts from approximately 2 to 10 days, depending on the pH of the surface to which it's applied; it breaks down faster on alkaline surfaces (or if mixed with alkaline water). Diazinon is moderately toxic to humans. It's more harmful to beneficials than malathion is, but not as deadly to them as carbaryl.

Carbaryl. A carbamate, this contact killer is effective against a broad range of pests, including many beetles and caterpillars. It lasts from about one to several weeks. As is true for diazinon, its lifespan is pH dependent. It's sold as a dust, liquid concentrate, or wettable powder.

continued on page 68

USING PESTICIDES SAFELY

The key to safe, effective pesticide use is to choose the right product for the targeted pest, then apply the material with the right equipment, at the right time. And *always* treat pesticides with respect, since even relatively nontoxic products can cause damage if used carelessly.

Follow these rules when choosing and applying pesticides.

■ Before you start thinking about which pesticide to use, ask yourself if you've exhausted all other possible forms of control. If you see natural enemies at work, give them a chance to catch up with the pest before you turn to chemical weapons. *Use pesticides only as a last resort.*

■ Select the right pesticide for the job. That means properly identifying the pest and the afflicted plant, then making sure the product is approved for use on both. Applying the wrong pesticide wastes time, effort, and money, and it may be harmful as well.

■ If more than one pesticide will do the trick, opt for the least toxic choice. Also look for selective products (those which target a few pests rather than a broad range) and products that quickly break down into harmless substances.

■ Buy only the amount you can use in one season. Don't be swayed by sales or economy-size packages unless you can share the product.

■ Read the label and follow the directions exactly. Use only the recommended amount; don't assume that if a little is good, a lot is better.

■ Time your treatments to catch pests at their most vulnerable stage.

■ If possible, spot-treat: use pesticides only in problem areas, not on the whole garden.

If you have any questions about the safety, environmental impact, or proper use of a pesticide, consult your state pesticide agency, your regional EPA office, or the National Pesticide Telecommunications Network, a 24-hour telephone hotline (800-858-PEST) funded by the EPA and Texas Tech University Health Sciences Center School of Medicine.

Reading the Label

This basic rule about pesticide use can't be repeated often enough: *read the label.* Read the label before

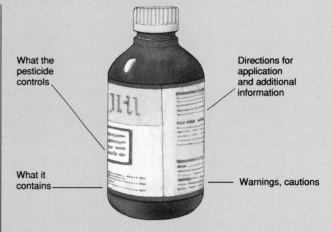

What the pesticide controls

What it contains

Directions for application and additional information

Warnings, cautions

buying a product, before mixing it, before applying it, before storing it, and before throwing it away. Don't trust your memory—read the label at every step.

The product label is the most important source of information about the pesticide. It's a legal document that tells you a product's effect on human health and the environment; its active ingredients; the pests and crops on which it can successfully be used; how to mix and apply it; whether it can be blended with other materials; whether there's a waiting period between applying the pesticide and harvesting crops or reentering the area; and how to store and dispose of the product (and its container). The label also spells out any special safety measures you'll need to take.

If you're in doubt about a product's suitability for your situation, the label is the absolute law. Other recommendations, such as those listed in the rogues' gallery starting on page 71, are intended as a guide.

Understanding the Risks

A signal word printed in large letters indicates the relative toxicity of a chemical. "Danger" or "poison" means highly toxic; "warning" means moderately toxic; and "caution" means slightly toxic. Relatively nontoxic materials don't require a signal word. Different formulations of the same pesticide often have different signal words: a more concentrated version may rate "warning," a more dilute one "caution."

Besides the signal word, product labels also include some text that details risks and notes safety precautions. You may see statements like the following: "Causes irreversible eye damage." "Wear goggles or face shields and rubber gloves when handling." "Do not breathe spray mist." "Wash skin thoroughly with soap and water after handling."

The days-to-harvest figure stated for pesticides that can be used on edible plants gives you a clue about that product's effect on humans. The longer the required interval from treatment to safe harvest, the more dangerous the pesticide.

Relative Toxicity of Some Pesticides

Scientists calculate toxicity by LD_{50} ratings, which refer to the amount of material needed to kill half of the test population. These numbers don't appear on pesticide labels, but are used in determining the signal word. The higher the number, the less toxic the pesticide.

The chart below gives ratings for some of the products mentioned in this book.

Insecticidal soap	16,500	Relatively
Bt	15,000	nontoxic
Neem	13,000	(no signal word)
Sabadilla	4,000	
Pyrethrum	1,500	
Diazinon	1,300	
Ryania	1,200	Slightly toxic
Malathion	1,000	(Caution)
Basic copper sulfate	1,000	
Acephate	866	
Carbaryl	850	
Bordeaux mixture	300	
Rotenone	132	Moderately toxic
Nicotine sulfate	55	(Warning)

Mixing & Applying Pesticides

Mix the pesticide in a well-ventilated area, using a separate set of measuring utensils labeled for pesticide use. Try not to mix more material than you can use in one application.

First put the water in a sprayer, then add pesticide—using the opposite order might leave the pesticide at the bottom. Pesticides usually work better if mixed with neutral or slightly acid water; if your water is alkaline, you can acidify it by adding a couple of teaspoons of white vinegar for each gallon of spray. Add the vinegar to the water and swirl it around before pouring in the pesticide.

When handling pesticides, wear protective gear as recommended on the product label, and don't smoke, eat, or drink. Before beginning the treatment, remove pets and toys from the area; also cover birdbaths, fish ponds, and anything else you don't want sprayed or dusted.

Apply pesticides only in still weather. When the air is calm, the material won't drift back onto you or onto plants that you didn't intend to treat. Coat the plant thoroughly, paying special attention to the leaf undersides, where many pests feed.

If you're using a pesticide harmful to bees, apply it late in the day, when bees are less active. If possible, use a spray rather than a dust, since bees are more likely to pick up dusts on their bodies.

Spraying. To prevent injury to your plants, water them thoroughly before spraying—thirsty plants can experience even more water stress when exposed to soaps, oils, or solvents in a pesticide formulation. Let the foliage dry, then apply the pesticide until it just begins to drip off the leaves.

A ready-to-use spray in an aerosol can or trigger sprayer is convenient for a small job. In all other cases, it's more economical to use liquid concentrates and your own sprayer. Good-quality plastic sprayers are an excellent choice: they're lightweight and corrosion-resistant, and it's easy to see how much solution you have.

A hand-held trigger sprayer is fine for a few plants or for spot-treating, but squeezing the trigger can be tiring if you're applying pesticide on a large

Using a backpack sprayer

scale. For big jobs, a pressurized tank provides the most precise application. Most home gardeners use hand-held models with capacities of 1 or 2 gallons. If you plan to spray a very big garden or a small orchard, you may want to invest in a backpack model. You need a strong back for this apparatus, which holds about 4 gallons of solution.

Another way to get pesticide into a small tree is with an old-fashioned trombone sprayer. You place the hose end into an open bucket of spray mixture, then slide a mechanism near the nozzle end to shoot the pesticide 10 to 20 feet into the tree.

Hose-end sprayers apply spray quickly, but they're not very precise in their calibration or in the way they distribute pesticide. They're most often used for spraying lawns. You put the pesticide in the sprayer, then attach the garden hose; as water runs through the hose, it mixes with the pesticide (the dilution rate is fixed on some models, variable on others). Get a model with an on-off switch that can be turned off independently from the water supply. The sprayer or hose bib should have a backflow preventer so that pesticide won't be siphoned into your water system if the water pressure drops.

Thoroughly clean your equipment after each use. Rinse several times with clean water, then operate the sprayer until clear water runs out of the nozzle. Keep a separate sprayer for weed killers, since any herbicide residue can harm plants that are later treated with other types of pesticides.

Dusting. Dusts adhere better to wet foliage, so mist plants before treating them—but apply dust during a dry spell, or rain will wash away the pesticide. Apply a thin coat; don't layer the material on. Since dusts are irritants, *always* wear a dust mask, even if you're handling the material for just a minute or two.

Some products come in ready-to-use squeeze bottles, but in most cases you'll need your own applicator. One type of duster has a bellowslike body that you squeeze, forcing the dust out through a nozzle. This device is tiring to use for all but the smallest jobs.

A more common applicator consists of a long tube with a pump that slides to propel a fine stream of dust through the nozzle. Another type of duster, better suited for large jobs, has a handle you crank as you walk along; it delivers a cloud of dust.

Storing Pesticides

Don't keep pesticides in the house. Store them in their original containers in a cool, dark, dry place,

Using a duster

such as a locked cupboard in a garage or shaded storage shed (not a metal shed). Avoid areas experiencing extreme fluctuations in temperature: the ideal range is 50° to 75°F. For details on proper storage of specific chemicals, follow label directions.

Most chemical companies formulate their products to last a minimum of 2 years in the container (exceptions are pesticides made from living organisms like *Bt*), although they can remain effective much longer under ideal storage conditions.

Pesticides don't lose their pest-control powers all at once. Instead of being 100 percent effective after several years, they may perform at only 70 to 80 percent of their original potency. The only reliable way to determine whether a pesticide will still do the job is simply to use it. Even if you write the purchase date on the container, you don't know how long the product sat in the store before you bought it.

Getting Rid of Leftover Pesticides

As long as a product hasn't been banned or restricted (limited to certain plants or pests or to professional application), it's environmentally safer to use it up as needed than to throw it away. However, you shouldn't use a product just to get rid of it. Apply it only if there's a problem, and only on the plants and pests listed on the label. If you don't want the pesticide yourself, give it to someone who does. Never dump it onto the soil or pour it down a drain.

If your cupboard holds any banned products, such as those containing chlordane, DDT, lead, or mercury, discard them according to local ordinances. For information on disposal in your area, call your local sanitation service, fire department, or state pesticide agency.

Although only slightly toxic to humans, carbaryl is deadly to bees and should not be sprayed on plants in bloom. And because the pesticide is toxic to creatures that prey on mites, spider mite infestations often follow carbaryl use. Bait forms suitable for pests such as cutworms and grasshoppers are less damaging, since only the intended victim is likely to be affected.

Acephate. This organophosphate is a systemic insecticide—it's absorbed into a plant, so that insects feeding on either the outer surface or the sap will be poisoned. Use it only on ornamentals, not on edibles. Acephate remains active for 6 to 9 days and is toxic to bees. It's typically sold as a liquid concentrate, but there's also a cartridge form, which is injected into trees and doesn't harm beneficials.

Fungicides

Though they're highly effective, synthetic fungicides are controversial choices for disease control. Many have been taken off the market due to concerns about carcinogenicity; yet others have been withdrawn by the manufacturers for various reasons. Many are still available, but for restricted use only.

Many of the synthetic fungicides sold today just coat the outside of the plant and act as a protective barrier. They prevent diseases but do not heal infected tissue. A few are systemic—they're absorbed by the plant and can cure infections. Consult your Cooperative Extension agent or experienced nursery personnel for more on currently marketed products.

The tomato plant on the right was genetically engineered; it produces a Bt *toxin that gives it resistance to hornworms. The plant on the left, stripped of its foliage by the caterpillar pest, doesn't contain the gene.*

FRONTIERS OF PEST CONTROL

In the past, pest-control research involved producing a steady stream of new chemicals. Today, however, investigators are setting out in largely unexplored directions. They're manipulating chemicals to make them safer, experimenting with gene alteration, and harnessing viruses and other organisms.

"Chemical prospecting" is one of the most exciting areas of research. Scientists are racing to explore the world's rainforests, hoping to learn more about the natural chemicals that these regions' multitudinous plants and insects use for self-defense. The ultimate goal is to mass-produce these chemicals or adapt them for use in growing fields and gardens.

Scientists are also stepping up work on viral pesticides—genetically altered versions of naturally occurring viruses that kill certain pests (cabbage loopers, for example) without harming other organisms. Many such pesticides are expected to be available to home gardeners in the future.

Another area of concentration is encapsulation—placing pesticides inside tiny capsules. These "pest pills" are intended to lengthen the life of the pesticide by protecting it from sunlight, rain, and other factors hastening its disintegration. A pathogen that ordinarily breaks down in days might last as long as several weeks if encapsulated. The capsule itself may be a starch coating or, in the case of some *Bt* products, the more stable cell wall of another organism.

Encapsulation can also make chemical pesticides safer to use. Because the capsule releases a product slowly, in a specific location, it's possible to use less than would otherwise be required—and with less contamination of ground water.

Genetic engineers are developing plants that produce their own pesticides: any hapless insect that stops to dine on leaf, stem, or flower will die. Efforts to splice *Bt* genes into plants have sparked controversy, however, since some scientists fear that the end result might be *Bt*-resistant insects. Researchers hope to overcome that risk by alternating *Bt* strains, or by arranging for the toxin to be triggered only in specific plant tissues or only when tissue is breached.

Pest-control experts are pinning many of their hopes on growth regulators like neem (see page 63). These pesticides interfere with an insect's biochemistry, preventing pests from maturing and reproducing. Growth regulators efficiently dispatch insects while posing minimum risk to humans and other mammals.

MAIL-ORDER SUPPLIERS & SOURCES OF INFORMATION

Local nurseries and garden centers are excellent sources of pest control supplies and information—but if you can't find what you're looking for, try the following mail-order companies and organizations. They're among the growing number of sources available to home gardeners, and you may find still other helpful groups and companies in your area. The addresses and telephone numbers listed here are accurate as of press time. Some of the suppliers provide catalogs free, while others charge a nominal fee.

MAIL-ORDER SUPPLIERS

Arbico
Box 4247 CRB
Tucson, AZ 85738
(800) 827-2847
(Row covers, pheromone and other traps, hand lenses, beneficial organisms, natural chemical controls, sprayers and dusters, soil care products)

Bountiful Gardens
18001 Shafer Ranch Road
Willits, CA 95490
(Row covers, netting, copper barriers, beneficial insects, natural chemical controls)

Bozeman Bio-Tech
1612 Gold Avenue
PO Box 3146
Bozeman, MT 59772
(800) 289-6656
(Pheromone and other traps, beneficial organisms, natural chemical controls)

California Environmental Protection Agency
Department of Pesticide Regulation
Environmental Monitoring and Pest Management
1220 N Street, Room A-149
PO Box 942871
Sacramento, CA 94271
(916) 654-1144
(For suppliers of beneficial organisms)

Gardener's Supply Company
128 Intervale Road
Burlington, VT 05401
(800) 444-6417
(Traps, row covers, electric fences, parasitic nematodes, natural chemical controls, sprayers and dusters)

Gardens Alive!
5100 Schenley Place
Lawrenceburg, IN 47025
(812) 537-8650
(Row covers, pheromone and other traps, repellents, beneficial organisms, natural chemical controls, sprayers and dusters)

Great Lakes IPM
10220 Church Road NE
Vestaburg, MI 48891
(517) 268-5693
(Pheromone and other traps, beneficial organisms, row covers, netting, hand lenses)

Harmony Farm Supply & Nursery
PO Box 460
Graton, CA 95444
(707) 823-9125
(Row covers, netting, copper barriers, repellents, pheromone and other traps, beneficial organisms, natural chemical controls, sprayers and dusters)

IFM (Integrated Fertility Management)
333 Ohme Gardens Road
Wenatchee, WA 98801
(800) 332-3179
(Row covers, pheromone and other traps, beneficial organisms, natural chemical controls, fertilizers, soil care products)

Necessary Trading Company
1 Nature's Way
New Castle, VA 24127
(800) 447-5354
(Row covers, pheromone and other traps, hand lenses, beneficial organisms, natural chemical controls, sprayers and dusters, soil care products)

Nitron Industries, Inc.
PO Box 1447
Fayetteville, AR 72702
(800) 835-0123
(Fertilizers, soil care products, beneficial organisms)

Peaceful Valley Farm Supply
PO Box 2209
Grass Valley, CA 95945
(916) 272-4769
(Row covers, netting, pheromone and other traps, repellents, hand lenses, beneficial organisms, natural chemical controls, sprayers and dusters)

Ringer Corp.
9959 Valley View Road
Eden Prairie, MN 55344
(800) 654-1047
(Natural chemical controls, fertilizers, soil care products)

Sterling International, Inc.
PO Box 220
Liberty Lake, WA 99019
(800) 666-6766
(Insect traps, spined soldier bug lures)

INFORMATION

Bio-Integral Resource Center
PO Box 7414
Berkeley, CA 94707
(510) 524-2567
(Membership includes subscription to "Common Sense Pest Control Quarterly")

California Environmental Protection Agency
See MAIL-ORDER SUPPLIERS
(Free 32-page booklet, "Suppliers of Beneficial Organisms in North America")

Cooperative Extension Offices
Look in the government pages of your telephone directory under "Government Offices, County" for "Cooperative Extension" or "Agricultural Extension"
(Advice and publications on a variety of topics)

National Pesticide Telecommunications Network
(800) 858-PEST
(24-hour telephone hotline)

An eastern fox squirrel nibbles on an ear of corn.

ROGUES' GALLERY

For help in identifying the mischief makers in your garden, look to this chapter. You'll find profiles of over 100 pests and diseases; some are familiar to gardeners everywhere, while others confine their dirty deeds to specific regions. Some strike a broad range of plants; others attack only one or a few types. Your gardening practices, too, play a role in determining the troublemakers you'll encounter.

The profiles are divided into three sections: insects and their relatives, larger creatures, and plant diseases. Each section contains an alphabetical listing of pesky organisms, including useful information about each: its looks, its *modus operandi,* and the control measures proven effective against it.

To exclude cases of mistaken identity, start by reviewing the list of beneficial creatures on pages 36–39. Make sure that your "pest" really is an evildoer—and not a garden ally.

Once you're convinced you have a scoundrel on your hands, skim the pest entries. Or, to save time, check the list of common plantings and the pests that attack them (pages 106–110). For example, if something is eating your rosebushes, look up "Rose" in the chart under "Trees & Shrubs." Then refer to the information on each listed pest.

You may recognize the villain from its mug shot or from the description of damage; perhaps you've noticed one of the "calling cards" mentioned for a particular culprit, such as characteristic holes chewed in leaves. The malefactor's usual active period—night, daylight hours, or a certain time of year—may provide another clue.

Be careful about fingering a suspect solely from its photograph—make sure the accompanying information checks out. Many creatures resemble each other, and the one you've encountered may not even be profiled in this chapter. Your Cooperative Extension office or a good local nursery can assist you in resolving any uncertainties; other gardeners in your area may also be helpful.

When you're fairly sure you've correctly identified the trouble, review the controls listed for that particular pest. In most cases, you can choose from nontoxic as well as more lethal remedies. The chemical controls we cite are suggestions only; before use, always read the product label to make sure the material is appropriate for your situation and for the plant and pest at issue.

Only the controls likely to work for most gardeners are noted, but other remedies are entirely possible. To learn more about the controls mentioned, consult "Preventing Problems" (pages 21–39) and "Dealing with Pests & Diseases" (pages 41–69).

Of all the pests that make themselves at home in gardens, the great majority are six-legged creatures and their kin. The following pages provide information on more than five dozen common troublemakers, some of which you'll doubtless have to combat at one time or another.

When you read about a particular pest, keep in mind that its feeding habits and life cycle are especially helpful in telling you which controls to launch—and when. For insects that undergo complete metamorphosis, we've indicated which controls (if any) are appropriate for each life-cycle stage.

ANTS

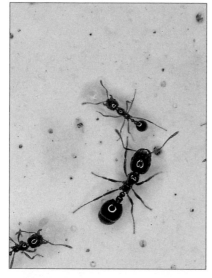

Many kinds of ants live in highly organized colonies in the soil, under rocks, and in tree cavities. Most types don't eat plants, but instead cause trouble by driving away creatures that prey on or parasitize sap-feeding pests such as aphids, whiteflies, mealybugs, and soft scales. All these pests excrete honeydew, a sugary sap ants like to eat. A column of ants marching up and down a tree trunk usually indicates an infestation by sap feeders.

Fire ants, a major pest in the South, do feed on plants—as well as on other insects. They build hard mounds up to 2 feet tall and inflict painful stings when disturbed.

Don't confuse ants with termites; ants are distinguished by their narrow waists and elbowed antennae.

Target: Honeydew (most species); fruits and vegetables (fire ants).

Damage: Most ants don't damage plants directly. Fire ants are an exception; they devour plants and spoil gardens with their mounds.

Life cycle: Winged males and females hatch about once a year. After mating, the males die and the females lose their wings. Each female establishes a nest, laying large numbers of eggs that hatch into workers. Colonies overwinter in soil or garden debris.

Control: Sticky barriers, diatomaceous earth, soap sprays, avermectin, boric acid, pyrethrum, diazinon, carbaryl.

Notes: Control ants to give sap feeders' natural enemies a better chance of suppressing their targets. The microbial pesticide avermectin prevents the queen ant from laying eggs, causing the colony to die out.

APHIDS

Small, pear-shaped, and soft-bodied, these slow-moving sucking insects usually congregate in clusters. They're sitting ducks for many natural enemies—but nonetheless, they prosper, thanks to their prolific reproduction. The many species come in a wide range of colors; usually wingless, they're equipped with two tubes, called cornicles, at the posterior end.

Target: Any soft plant tissue, including new growth of most woody plants.

Damage: Juices are sucked from tender growth, causing wilting or malformations. Honeydew—the undigested plant sap that aphids excrete—attracts ants and fosters sooty mold (see page 104). Some aphids transmit viruses.

Life cycle: Aphids multiply without mating for most of the year, producing live young that reproduce within a week or so after birth. Some species give birth to pregnant young. Breeding is continuous in mild climates; in colder regions, there's a winged form that deposits eggs in fall.

Control: Handpicking, pruning, water jets, aluminum-foil mulch, yellow sticky traps, aphid midges, green lacewings, soap sprays, oil sprays, diatomaceous earth, sulfur, pyrethrum, neem, diazinon, malathion, acephate.

Notes: Use sticky traps for winged aphids. Avoid persistent chemicals if you want help from natural enemies such as ladybird beetles, green lacewings, syrphid flies, and parasitic wasps (hollow aphid bodies with a round exit hole indicate that wasps are on the job). Control ants (see above), which protect aphids from natural enemies.

ARMYWORMS

Active at night and on overcast days, armyworms get their name from their habit of marching in troops, devouring vegetation in their path. Several species of these hairless caterpillars cause headaches for gardeners; all are 1 to 2 inches long and fond of leafy vegetables. Beet armyworms, varying from green to black, sport a stripe down each side and a small black spot above the stripe at the middle pair of true legs. Yellow-striped armyworms are purplish to black, with two yellow stripes running along their backs. Brown, shiny fall armyworms have prominent black spots and a white "Y" on a black head. (Armyworms are also known as climbing cutworms; see page 78.)

Target: Vegetables, especially seedlings (fall armyworms favor corn); lawns.

Damage: Foliage is chewed (tips of corn leaves look ragged as they unfurl); seedlings are often destroyed. Grass dies in irregular patches.

Life cycle: Moths lay eggs on leaves. After hatching, the caterpillars feed for several weeks, then pupate in the soil. Beet and fall armyworms migrate north, reaching the upper part of their range in autumn. Depending on species and climate, there are one to six generations each year.

Control: Trichogramma wasps (eggs); handpicking, weeding, ditches, *Bt,* parasitic nematodes, spined soldier bugs, neem, diazinon, carbaryl (larvae); tilling (pupae).

Notes: Get rid of grassy weeds, since the infestation often starts there. Dig a steep-sided ditch at least 6 inches deep around the plantings you want to protect; the armyworms will fall in, then be unable to get out.

ASPARAGUS BEETLES

As the name implies, these pests—both adults and larvae—prefer asparagus to all other foods. The adults, reaching about ¼ inch long, are blue-black mottled with yellowish orange. The plump, ⅓-inch-long grubs are olive green to gray, with a black head and legs.

Target: Asparagus.

Damage: Growing tips of spears are chewed and scarred.

Life cycle: When shoots appear in spring, adults fly in to feed, then lay dark eggs that protrude horizontally from spear tips. The grubs hatch out within a week, feed for a few weeks, and then burrow into the ground to form yellow pupae. Adults can overwinter in plant debris. There are two to five generations a year.

Control: Row covers, sanitation (adults); prompt harvesting, handpicking, water jets, rotenone, malathion, carbaryl (adults and larvae); spined soldier bugs (larvae).

Notes: Many natural enemies will help you keep this pest in check.

BAGWORMS

The dangling silken bags, reaching a length of 2 inches, are more noticeable than the small brown caterpillars—each of which weaves its own bag and drags it along as it feeds. Bagworms are found as far west as Texas.

Target: All trees, especially juniper and arborvitae.

Damage: Leaves are chewed from the treetop down; defoliated trees may die. Deciduous trees are less vulnerable, since they can grow a new set of leaves. The silken bands attaching the bags to trees may girdle and kill young branches.

Life cycle: In winter, each bag contains as many as 1,000 eggs. In spring, the caterpillars hatch and disperse to feed for a time; then each one tethers its bag to a twig and enters to pupate. In a few days, the moths emerge. The females, wingless and mouthless, remain in their bags; the small, black-winged males fly in to join them. After mating and laying eggs, the females die. There is one generation a year.

Control: Pheromone traps (adults); handpicking bags (eggs); *Bt,* malathion, diazinon, carbaryl, acephate (larvae).

Notes: Pheromone traps set out in summer help reduce the next year's population. Treat the larvae as soon as they hatch, before they construct their bags.

BARK BEETLES & BORERS

In their larval stage, numerous types of beetles and some clearwing moths tunnel beneath bark or bore into live wood. Some borers tunnel deep into a branch, making it weak enough to snap in a storm; others tunnel just below the bark, girdling a tree. In many cases, a pest's tunneling pattern is distinctive enough to provide positive identification.

Borers tend to attack trees stressed by poor growing conditions or wounds. If you see holes bordered with sawdust, excrement, or sap, you probably have an infestation on your hands. Some species attack below the soil line; in these cases, the evidence may be at the base of the tree.

Target: Many trees and shrubs.

Damage: Wilting, yellowing, premature leaf drop, and branch dieback occur. Severely infested plants may die: tunneling provides an entrée for infections, and bark beetles themselves often spread diseases as they tunnel.

Life cycle: Bark beetles bore beneath bark and lay eggs in tunnels; moths lay eggs around wounds and on or at the base of host plants. The larvae of both kinds of insects tunnel, feed, and pupate inside the plant, then emerge as adults. Most of these pests overwinter in a dormant state.

Control: Hand worming, pruning and burning infested wood, injecting parasitic nematodes into holes.

Notes: Prevention is the best control. Provide good growing conditions and place barriers around trees to protect them from injury. To hand worm, ram wire into borer holes to kill the occupants. Clean out the holes before injecting nematodes.

BILLBUGS

A long, forward-pointing snout gives this dark, slow-moving, ½-inch-long weevil its name. The real pests, though, are the fat, legless grubs. About the same size as the adult beetles, they're white with brown heads. Billbugs are a problem primarily on the East Coast and in the West.

Target: Lawns, especially those planted with Bermuda or zoysia grass.

Damage: Irregular brown patches that lift right up from the lawn appear where grass roots have been chewed through.

Life cycle: Adults hibernate in protected areas, then lay eggs in grass stems in spring. The grubs hatch out and tunnel into the stems, then move deeper to feed on crowns and roots before pupating in autumn. There is one generation a year.

Control: Parasitic nematodes, diazinon.

Notes: The legless larvae offer the best evidence of a billbug infestation; other lawn pests that cause similar damage have legs. You'll also see whitish, sawdustlike excrement in infested areas.

BLISTER BEETLES

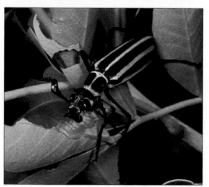

Although markings vary greatly depending on the species, blister beetles are easily identified by their physique: the middle section of the body is noticeably smaller than the head or abdomen. The majority of these pests are well under an inch long.

Target: Many flowers and vegetables, especially those belonging to the tomato family.

Damage: Leaves are chewed. When large populations of beetles appear suddenly, damage can be serious.

Life cycle: Adults lay eggs beneath the soil, where the newly hatched grubs feed on grasshopper eggs. The grubs remain underground in a dormant state for up to 2 years, then become active, pupate, and emerge as plant-feeding adults in late spring or early summer.

Control: Handpicking, rotenone, pyrethrum, carbaryl (adults); tilling, parasitic nematodes (larvae and pupae).

Notes: Use gloves when handpicking these beetles, since they exude a substance that can cause painful blisters.

CABBAGE LOOPERS

Green with pale stripes, this caterpillar has legs at the front and back, but none in the middle—a peculiarity that gives it a characteristic humpback or looping crawl. Fully grown, the pests are about 1½ inches long.

Target: Cole crops, lettuce, celery, tomatoes, and various other vegetables and flowers.

Damage: Large, irregular holes are chewed in leaves. Holes are bored in tomatoes and in cabbage and lettuce heads. Seedlings may be killed.

Life cycle: Moths lay greenish white eggs singly on leaves. The caterpillars feed for a few weeks, then pupate on the plant; pupae may overwinter, or adults may migrate south. There are two to seven generations a year.

Control: Row covers, pheromone traps (adults); trichogramma wasps (eggs); handpicking, *Bt*, green lacewings, spined soldier bugs, pyrethrum, rotenone, neem, carbaryl (larvae).

Notes: Wasps parasitize cabbage loopers; if you see eggs on caterpillars, leave those loopers to their fate. Also leave chalky white loopers alone—they're infected with a virus. Since cabbage loopers feed first on outer leaves, you can often just pull the marred leaves from heads of cabbage or lettuce and still get a good harvest.

CANKERWORMS

Also called inchworms for their looping motion, these 1-inch-long pests are found in all but the southernmost parts of the country. Spring cankerworm moths show up in spring, fall cankerworm moths in autumn when weather turns cold. The caterpillars of both types feed in spring, moving from tree to tree on silken strands. Spring cankerworms are green, brown, or black with colored stripes, while the fall type is brown on top and green below, with three slender white stripes.

Target: Fruit and shade trees.

Damage: Leaves are chewed, but no webbing is visible. Trees may be defoliated during a bad infestation. Young fruit may have holes that heal over into scabby patches.

Life cycle: Wingless female moths crawl up tree trunks and lay eggs. The spring species lays clusters of oval, purplish eggs beneath the bark; the fall type deposits neat tiers of gray, flowerpot-shaped eggs on twigs. The eggs hatch when trees leaf out in spring; the larvae feed for about a month, then lower themselves by silken threads and form underground pupae. There is one generation a year.

Control: Sticky bands (female moths); oil sprays, trichogramma wasps (eggs); *Bt*, spined soldier bugs, neem, malathion, carbaryl, acephate (larvae); tilling (pupae).

Notes: Heavy infestations usually occur every 7 years or so. Repeated defoliation weakens a tree.

CHINCH BUGS

This true bug sucks sap from grasses; it's attracted to poorly grown lawns. Chinch bugs start out pinhead-size and bright red, with a white band across the back; they darken as they mature, eventually becoming black, ⅛-inch-long bugs with white wings. Don't confuse this pest with its natural enemy, the predaceous big-eyed bug: the faster-moving predator is wider and has prominent eyes.

Target: Lawn grasses, corn, and other grasses.

Damage: Afflicted plants wither and dry out. Dead lawn remains firmly rooted.

Life cycle: Adults overwinter in tall grass or debris. When the weather warms in spring, they lay eggs on the grass or in the soil. There are one to seven generations a year.

Control: Resistant varieties, dethatching, soap sprays, sabadilla, pyrethrum, diazinon, carbaryl.

Notes: To see if chinch bugs are present, remove both ends from a large can; then push it into the ground and fill it with water. Let stand for 10 minutes, then check—the bugs will have floated to the top. Before treating, water the lawn to bring bugs to the surface. No treatment is needed if the bugs you see are coated with a gray, cottony material—they're infected with a fungus.

CICADAS

These 1- to 2-inch-long aphid relatives have a black or mottled body, prominent reddish eyes, short antennae, and transparent wings. Periodical cicadas appear in late spring or early summer; the less harmful dogday cicadas show up in mid to late summer. Male insects produce a loud, buzzing "song" by vibrating membranes on the underside of the abdomen.

Target: Many trees and shrubs, especially apple, peach, oak, and dogwood.

Damage: Feeding by the insects usually doesn't do much harm, but slits made for egg-laying cause twig dieback.

Life cycle: These pests spend most of their lives below ground as nymphs, feeding on tree and lawn roots. Each brood of dogday cicadas spends 2 to 4 years developing underground; periodical cicadas spend about 13 years below ground in the South, 17 years in the North. Once the nymphs dig their way out, they climb into trees and molt for the last time. The adults live for several weeks, during which time they mate and lay eggs in twigs. After about 2 months, the eggs hatch; the nymphs drop to the soil and tunnel down.

Control: Sticky bands, netting, pruning, parasitic nematodes, carbaryl.

Notes: Since broods overlap, periodical cicadas may emerge more often than once every 13 or 17 years. Don't plant new trees until after egg laying; your Cooperative Extension office can tell you when a new brood is expected. Netting on young trees should have ¼-inch-diameter or smaller holes.

CODLING MOTHS

This 1-inch-long caterpillar, with a white or pinkish body and a dark head, is one of the most stubborn orchard pests. The damage is worse during warm, dry springs.

Target: Fruit and nut trees, especially apple, pear, and walnut.

Damage: Fruit is tunneled and often drops prematurely; the caterpillar may still be inside.

Life cycle: Coppery-banded moths lay eggs in spring when host trees are in bloom. After hatching, caterpillars tunnel to the core of the developing fruit or nut. They eventually burrow out, leaving excrement, then drop to the ground or crawl down the trunk to pupate in cocoons under loose bark or in debris. There are two to four generations a year.

Control: Pheromone traps (adults); oil sprays, trichogramma wasps (eggs); destroying fallen fruit, corrugated cardboard collars, *Bt,* ryania, malathion, carbaryl (larvae).

Notes: Bt is effective only during the brief period before larvae enter fruit. A collar is the best means of trapping overwintering larvae. To ensure that the pests use the collar, make it the most attractive choice: scrape the bark smooth, so the larvae won't pupate under bark flaps.

COLORADO POTATO BEETLES

Though native to the Rocky Mountains, this pest long ago spread to all other parts of the United States except California and Nevada. The ⅜-inch-long adult beetle is easy to spot: the showy polka-dot vest and striped pants are dead giveaways. The small, red-humped grubs have articulated legs and rows of dark spots along both sides. Both adult beetles and grubs denude plants, leaving black excrement as testament to their gluttony.

Target: Tomato-family vegetables and flowers.

Damage: Leaves and stems are chewed; whole plants may be devoured if the beetle population is large.

Life cycle: In spring, female beetles lay hundreds of elongated orange eggs in clusters on leaf undersides. About a week after egg laying, the larvae emerge and feed for a time, then burrow into the ground to pupate. The pupae can overwinter. There are one to three generations each year.

Control: Resistant varieties, crop rotation, timed plantings, row covers, handpicking, thick mulch, rotenone (adults); *Bt san diego,* spined soldier bugs (larvae); tilling, parasitic nematodes (pupae).

Notes: A thick mulch of straw or hay slows down the beetles. The insect is resistant to most pesticides.

CORN EARWORMS

Also known as the tomato fruitworm and cotton bollworm, this caterpillar changes appearance markedly as it grows. Young corn earworms are tiny and white, with black heads; older ones are 1½ inches long and green to nearly black in color, with lengthwise stripes and stubby spines along their backs. Finding one of these pests in an ear of corn is dismaying—but the damage is modest, and the ear can be salvaged by cutting off the spoiled tip.

Target: Corn, tomatoes, and other vegetables.

Damage: Caterpillars and eggs are found in silk and tip kernels of ripening corn; tomatoes may be tunneled.

Life cycle: In spring, moths lay domed, ridged, whitish eggs singly on silks or leaf undersides. Caterpillars hatch and feed for several weeks, then pupate in the soil. There are up to seven generations a year.

Control: Delayed planting, pheromone traps (adults); trichogramma wasps (eggs); resistant varieties, handpicking, mineral oil, *Bt* granules, parasitic nematodes, spined soldier bugs, carbaryl (larvae); tilling (pupae).

Notes: Plant tight-husked corn varieties; or put a clothes pin or rubber band on tip of husk. Place a dropperful of mineral oil in tip of ear when silks have withered but before they turn brown. Apply *Bt* granules or inject nematodes into husk during silking.

CUCUMBER BEETLES

The various species of these ¼-inch-long beetles are yellowish green, with black spots or stripes. The slim, white larvae, about ½ inch long, are dark at both ends. Larvae feed on roots; adults—the more destructive stage—chew aboveground plant parts.

Target: Roots of corn, other grasses, legumes (larvae); many vegetables, especially cucurbits, and flowers (adults).

Damage: Large, roughly oval holes are chewed in leaves and flowers. As they feed, the beetles can spread bacterial wilt (see page 98) and a mosaic virus (see page 101).

Life cycle: When the weather warms in spring, beetles lay eggs beneath host plants, either at plant bases or in cracks in the soil. Larvae burrow into the soil and feed on roots, then pupate. Adults overwinter in debris. There are one to four generations a year.

Control: Resistant varieties, row covers, handpicking, rotenone, sabadilla, malathion, carbaryl (adults); parasitic nematodes, diazinon (larvae).

Notes: Try to catch infestations early; killing the beetles won't save your crop if the pests have already infected it with a disease.

CURCULIOS

These beetles resemble other members of the weevil group in having a long, curved snout. The larvae are grayish white, legless grubs. Most curculio species, including the plum curculio (shown), pecan weevil, black walnut curculio, and rose curculio, attack a narrow range of hosts. The plum curculio, found east of the Rockies except along the Gulf Coast, is among the most difficult-to-control pests of apples and stone fruits. Crescent-shaped or mushroom-shaped scars on fruit are its calling card.

Target: Fruit and nut trees, roses.

Damage: Fruits and nuts are tunneled. Infested fruit ripens and drops prematurely; nuts wither and fall. Holes are drilled in rose blossoms.

Life cycle: Adult beetles overwinter in debris, then feed on spring growth before laying eggs on developing fruits, nuts, or flower buds—into which the larvae tunnel to feed. After feeding, larvae burrow into the ground to pupate. There are one or two generations a year.

Control: Handpicking, malathion, carbaryl (adults); destroying infested plant parts (larvae); tilling, parasitic nematodes (pupae).

Notes: Adults play dead when disturbed, so you can shake them out of plants onto a cloth and dispose of them.

CUTWORMS

So named because they chew seedlings off at ground level, cutworms are the larvae of various moths. Up to 2 inches long and of diverse colors, the hairless caterpillars feed at night; during the day, they can be found underground or beneath debris near a food source, curled up in a C shape. (For information on climbing cutworms, see the discussion of armyworms on page 73.)

Target: Young seedlings and transplants; lawns.

Damage: Stems of young plants are chewed off near the ground. Leaves of older plants show ragged edges and chewed holes.

Life cycle: Moths lay tiny white eggs, often in weeds or debris. After feeding, the caterpillars pupate in the soil. The insect can overwinter at any stage of life. There are up to four generations each year.

Control: Weeding, trichogramma wasps (eggs); soil solarization, sanitation, cutworm collars, handpicking, tilling, parasitic nematodes, baited *Bt* granules, carbaryl bait, diazinon (larvae).

Notes: Work diazinon into the soil before planting. For a simple collar, remove both ends from an empty can, then place it over a seedling. Before handpicking, flood infested areas to force cutworms to the surface.

EARWIGS

Contrary to the old wives' tale, these nocturnal, ¾-inch-long, reddish brown insects don't crawl into human ears and bore into the brain with their mean-looking pincers. Earwigs do nibble on plants, though—but they usually do more good than harm, since they eat decaying matter and other insects. Due to their habit of hiding in gnawed fruit, they're often blamed for the evil deeds of other pests. Before you take action against earwigs, schedule a nighttime patrol to see if they're responsible for the damage you're finding.

Target: Tender plant tips.

Damage: Young growth is nibbled; leaves show ragged holes.

Life cycle: In fall, adults lay clusters of 20 to 60 pearly eggs in the top few inches of soil. The eggs hatch in spring. There's usually one generation a year, though there may be a second brood if some adults overwinter and lay eggs in summer.

Control: Handpicking, trapping, rotenone, carbaryl bait, diazinon.

Notes: Look for earwigs in cool, dark, damp places during the day. A loosely rolled newspaper makes a good trap.

ELM LEAF BEETLES

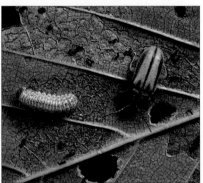

Both adults and larvae feed on leaf undersides, but the larvae—½-inch-long, dull yellow grubs with black stripes—cause more damage. The ¼-inch-long adults are yellowish green, with a black stripe on each wing cover.

Target: Elms, Japanese zelkova.

Damage: Leaves are chewed to lace; injured leaves may turn brown and drop. Repeated defoliation may kill a tree. Weakened trees are also susceptible to elm bark beetles, which spread Dutch elm disease (see page 100).

Life cycle: Adults overwinter in buildings and protected places. In spring, they fly to nearby elms to feed on unfolding leaves; they lay onion-shaped, yellowish eggs on the leaf undersides, then die. The larvae feed for several weeks, then crawl or drop down to the ground and pupate near the trunk base or in debris. There are up to five generations a year.

Control: *Bt san diego*, acephate (adults and larvae); spined soldier bugs, sticky bands, carbaryl (larvae); handpicking, sanitation (pupae).

Notes: Destroy elm leaf litter. Add summer oil to *Bt san diego* to kill eggs along with adults and larvae. Apply sticky material or spray a 2-foot-wide band of carbaryl high on the trunk to kill larvae as they crawl down.

EUROPEAN CORN BORERS

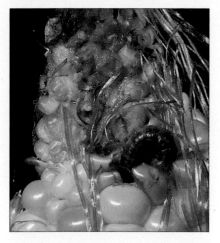

This 1-inch-long, grayish pink caterpillar has a dark head and two rows of dark spots along its body. Found everywhere except Florida and the Far West, it's most troublesome in the Midwest.

Target: Primarily corn; many other vegetables and flowers are also attacked.

Damage: Leaves show small "shot holes"; corn stalks have small holes edged with sawdustlike material. Tassels are broken; the stems attaching ears to the main stalk may be bent. Ears of corn may be tunneled.

Life cycle: Moths lay masses of white eggs on leaf undersides. After hatching, the caterpillars feed briefly on leaves and tassels, then tunnel into stalks and feed for several weeks. The caterpillars pupate (and may overwinter) in plant stems. There are one to three generations a year.

Control: Resistant varieties, weeding, pheromone traps (adults); trichogramma wasps (eggs); handpicking, baited *Bt* granules, ryania, rotenone, diazinon granules, carbaryl (larvae); destroying plant debris (pupae).

Notes: Weeded corn patches are less likely to be attacked than weedy ones. Place *Bt* or diazinon granules in leaf whorls on stalks. Make a short lengthwise slit in the stalk below the entrance hole and handpick the borer.

FALL WEBWORMS

About an inch long, this leaf-feeding pest is a long-haired, pale green or yellow caterpillar with a black stripe down its back. It feeds from inside the silken, weblike nests it builds at branch tips.

Target: Fruit, nut, and shade trees; roses and other ornamental shrubs.

Damage: Leaves are chewed; individual branches and sometimes entire plants are defoliated. Plants are covered with dirty white webbing.

Life cycle: White moths lay woolly egg masses on leaf undersides. After feeding, the caterpillars crawl down the trunk to pupate; pupae overwinter in litter or attached to tree bark. There are up to four generations a year.

Control: Trichogramma wasps (eggs); handpicking, pruning, sticky bands, spined soldier bugs, oil sprays, *Bt*, malathion, diazinon, carbaryl, acephate (larvae).

Notes: Use a long pole to destroy webs. When spraying, poke the sprayer through webs to get to the leaves the caterpillars are eating. Tent caterpillars (see page 91) are similar pests.

FLEA BEETLES

These tiny, oval jumping insects vary in color depending on the species, but most types are black, shiny bronze, or dark blue. Except for a desert species with a particular fondness for corn, all have an appetite for a broad range of edible plants. Adults chew holes in leaves; the white larvae feed on roots. Healthy older plants can survive a flea beetle attack, but seedlings may die.

Target: Many plants, especially dichondra and vegetable seedlings.

Damage: Irregular "shot holes" are chewed in leaves. The insects spread diseases as they feed.

Life cycle: Adults overwinter in weeds or garden debris. In late spring, they lay tiny white eggs just under the soil around host plants. Larvae feed underground, then pupate. There are one to four generations a year.

Control: Weeding, sanitation, row covers, white sticky traps, misting, diatomaceous earth, rotenone, carbaryl (adults); parasitic nematodes, tilling (larvae and pupae).

Notes: Since the pests prefer dry conditions, lightly mist plants under siege.

FRUIT FLIES

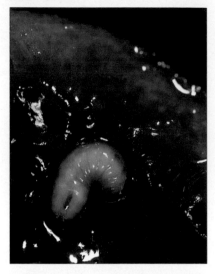

These hard-to-control orchard pests include the apple maggot, blueberry maggot, cherry fruit fly, walnut husk fly, and numerous quarantined exotic species (the Mediterranean fruit fly, for example). The native flies are a little smaller than houseflies; they're black with white stripes and have banded and spotted wings. The maggots, which feed inside fruits and nuts, are small, whitish, and legless.

Target: Many fruit and nut trees.

Damage: Fruits and nuts are decayed and wormy; they may rot on the tree or drop prematurely.

Life cycle: Adult flies lay eggs under the skin of fruits or nuts; the maggots burrow inside after hatching. Once fully grown, they eat their way out, drop to the ground (or crawl out of fallen fruit), and pupate in the soil. Native species have one generation a year.

Control: Sticky traps, copper, carbaryl (adults); destroying infested fruit (larvae).

Notes: Baited sticky traps are available for many species; for information on key control times, consult your Cooperative Extension office. Yellow sticky traps attract the majority of species. Apple maggot flies and some cherry fruit flies are also drawn to red spheres; green spheres attract walnut husk flies. New evidence indicates that copper compounds can reduce egg laying.

GERANIUM BUDWORMS

Also known as tobacco budworms, these close relatives of the corn earworm burrow into buds and feed from the inside; they also chew fully opened flowers. The striped, ¼- to ¾-inch-long caterpillars are greenish to tan or reddish in color; they cause problems largely in mild climates, since they can't survive cold winters.

Target: Primarily geraniums, petunias; also nicotiana, calendulas, border penstemon.

Damage: Flowers are tattered or don't bloom. Tiny droppings appear on buds.

Life cycle: Moths lay clusters of eggs on host plants. The caterpillars tunnel into developing buds and feed for a few weeks before pupating, usually inside the damaged buds. After emerging, the adults lay eggs that may either overwinter or hatch into a second generation.

Control: Resistant varieties (adults); oil sprays (eggs); *Bt*, pyrethrum, carbaryl, acephate (larvae).

Notes: Ivy geraniums are resistant. Apply *Bt*, pyrethrum, or carbaryl immediately after eggs hatch—the larvae are safe once they enter buds.

GRASSHOPPERS

Also called locusts, the dozens of grasshopper species differ in size, color, and markings—but all flourish in areas with long, hot, dry summers.

Target: All plants, especially grasses and weeds.

Damage: If the insects descend in hordes, they'll eat plants to the ground, but individual grasshoppers are a threat only to young plants.

Life cycle: The insects usually overwinter as yellowish eggs in the soil or on weeds. Nymphs hatch out early in spring and begin to feed, molting five or six times before reaching adulthood. Most species produce one generation a year.

Control: Row covers, handpicking, trapping, tilling, weeding, soap sprays, neem, malathion, diazinon, carbaryl, acephate.

Notes: In home gardens, row covers are probably the best control. To trap grasshoppers, fill jars with a solution of one part molasses and nine parts water, then bury them so the lip is at ground level. The pathogen *Nosema locustae* (see page 57) isn't effective in the average home garden.

GYPSY MOTHS

Since its accidental release in Massachusetts in 1869, this pest has defoliated millions of acres of trees in the East. A notorious hitchhiker (hence the name "gypsy"), it's gradually moving westward, traveling to new regions as egg masses attached to vehicles. An Asian strain which feeds on an even wider range of plants than the European type has shown up in the western United States.

Newly hatched caterpillars can disperse by hang-gliding on silken strands. Eventually growing up to 2 inches long, they're marked with rows of blue and red spots; tufts of fine hairs protrude from their sides. They feed in such numbers that their excrement rains upon the ground.

Target: Many shrubs and trees, especially oaks.

Damage: Leaves are eaten. Deciduous trees are only weakened, but defoliated evergreens may die. Damage is more severe when a warm spring follows a dry autumn.

Life cycle: Large tan moths lay chamoislike egg masses on any rough surface; the eggs overwinter and hatch in spring. The caterpillars climb host trees and feed on leaf undersides, then crawl down to form mahogany-brown pupae in bark crevices and on other surfaces. The moths soon emerge, mate, and die. There is one generation a year.

Control: Resistant plants, pheromone traps (adults); handpicking, trichogramma wasps (eggs); burlap or sticky bands, spined soldier bugs, *Bt*, neem, carbaryl, acephate (larvae).

Notes: Some people are allergic to gypsy-moth caterpillar hairs, so wear gloves when removing the pests from burlap or sticky bands. Gypsy moths build up to huge numbers every 5 to 10 years, but their numbers are then reduced by a fungal or viral disease.

HARLEQUIN BUGS

These handsome black-and-orange pests are found largely in the southern half of the country. Shield shaped and ¼ inch long, the harlequin is a type of stink bug, releasing a foul odor when disturbed. Its highly distinctive eggs resemble neat rows of tiny white barrels with black hoops. Both adults and nymphs suck plant sap.

Target: Primarily cole crops; many other plants are also attacked.

Damage: Yellowish patches appear on leaves. Heavy feeding can kill a plant.

Life cycle: Adults overwinter in debris, then lay eggs on leaf undersides. Nymphs feed for a few weeks after hatching. There are usually three or four generations annually, but in the Deep South, breeding can go on all year.

Control: Resistant varieties, row covers, sanitation, handpicking, trapping, soap sprays, sabadilla, rotenone, pyrethrum, carbaryl.

Notes: Handpick adults when they first appear in spring. To lure them away from crops, place cabbage leaves elsewhere in the garden; destroy the decoyed bugs daily.

HORNWORMS

Up to 5 inches long, hornworms are the larvae of large brown moths that fly like hummingbirds. Both tomato and tobacco hornworms are green with diagonal white stripes; the horn, at the pest's rear, is black on the tomato hornworm, red on the tobacco hornworm. The caterpillars feed upside down on leaf undersides. Since they blend into the foliage, they can be hard to spot, although the black pellets they excrete are plainly visible.

Target: Tomatoes, eggplant, potatoes, peppers, and other tomato-family plants.

Damage: Stems are stripped bare of leaves; fruit may be gnawed.

Life cycle: Moths lay pale green eggs singly on leaf undersides. After feeding, the caterpillars enter the soil and form brown, shiny, 2-inch-long pupae with a handlelike projection. The pupae can overwinter. There are one to four generations a year.

Control: Crop rotation, row covers (adults); trichogramma wasps (eggs); handpicking, *Bt* (larvae); tilling (pupae).

Notes: Braconid wasps parasitize hornworms. If you see worms with tiny white cocoons on their backs, let the wasps finish them off.

IMPORTED CABBAGEWORMS

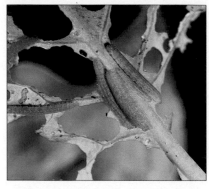

These velvety, light green caterpillars are similar to cabbage loopers (see page 75) in size, but more conventional in gait. Older caterpillars have faint yellow stripes.

Target: Primarily cole crops.

Damage: Large, irregular holes are chewed in leaves, and the parts that aren't eaten are contaminated with dark green excrement. Cabbage heads are tunneled.

Life cycle: White butterflies with black-tipped wings lay bullet-shaped, ridged yellow eggs on leaf undersides. After feeding for 2 weeks or so, the caterpillars pupate on plants, below ground, or on fences or other structures. The pest breeds all year long in mild regions, but may overwinter as pupae in cold climates. There are three to six generations a year.

Control: Row covers (adults); trichogramma wasps (eggs); handpicking, green lacewings, spined soldier bugs, *Bt,* pyrethrum, rotenone, neem, carbaryl (larvae).

Notes: When handpicking imported cabbageworms, look carefully; they're very difficult to spot.

JAPANESE BEETLES

Found primarily in the eastern United States, this pest eats almost everything except vegetables. The C-shaped grubs—whitish, up to 1 inch long, with brown heads and three pairs of legs—feed heavily on plant roots; lawns are especially likely to be attacked. When the ½-inch-long, metallic green beetles with coppery wing covers emerge, they decimate the aboveground parts of roses and other plants.

Target: Leaves, flowers, and fruit of over 275 plant species; lawns.

Damage: Leaves are skeletonized, and flowers and fruit are eaten (adults); brown patches of lawn roll up like carpet (larvae).

Life cycle: Grubs spend most of the year deep in the soil; in early spring, they come nearer to the surface to feed, then pupate. Adults emerge in summer, feed for about 6 weeks, and lay eggs in the soil. The cycle may take 2 years to complete.

Control: Handpicking, rotenone, pyrethrum, carbaryl, acephate (adults); milky spore disease, parasitic nematodes, tilling, diazinon granules (larvae).

Notes: Traps are risky, since they may attract more beetles than they can handle. Handpicking is probably more effective—the beetles play dead when disturbed, so they can be shaken onto a cloth and discarded. Some garden suppliers sell spiked sandals that do two jobs at once: you can impale grubs as you aerate your lawn.

JUNE BEETLES

Varying in color and size by the species, this beetle flies after dusk and is attracted to light. The various common names—May beetle, June bug, June beetle—refer to the season when adults are active, but it's the earlier larval form that should worry gardeners. Also called white grubs, the larvae resemble Japanese beetle grubs (see above) both in appearance and in the damage they do. Adults nibble on leaves of shade trees and roses, but their activities seldom cause problems.

Target: Roots of lawn grasses, strawberries, beans, beets, corn, onions, potatoes.

Damage: Roots are eaten; brown patches of lawn roll up.

Life cycle: Adults emerge in late spring and summer, then lay eggs in the soil in late summer. Grubs feed on roots until fall, then burrow deeper to overwinter. They move nearer to the surface in spring, resume feeding, and then pupate. The cycle can be completed in a single year in warm climates, but in colder regions, grubs may overwinter for 2 years, then pupate during the third winter.

Control: Handpicking, rotenone, pyrethrum, carbaryl (adults); parasitic nematodes, tilling, diazinon granules (larvae).

Notes: June beetle grubs are not susceptible to milky spore disease.

LACEBUGS

Various species of these true bugs are found in all parts of the country. Both the whitish, ⅛-inch-long, lacy-winged adults and the darker, wingless nymphs suck sap from leaf undersides. Despite their wings, adult lacebugs seldom fly; they have a slow sideways movement.

Target: Many ornamental trees and shrubs, especially azaleas and rhododendrons.

Damage: Leaves lose color and are speckled or blotched. Plants decrease in vigor and bloom poorly.

Life cycle: Adults insert egg clusters into leaf veins or cement them to leaves. Some species overwinter as eggs, others as adults. There are several generations a year.

Control: Water jets, soap sprays, oil sprays, pyrethrum, sabadilla, malathion, diazinon, carbaryl, acephate.

Notes: Unlike leafhoppers, which cause similar damage, lacebugs leave obvious spots of dark excrement on leaf undersides.

LEAFHOPPERS

There are some 2,500 species of these small, agile, wedge-shaped insects; many types are handsomely colored and patterned. Both adults and nymphs—which look like wingless or short-winged adults—suck sap from leaf undersides. Some species favor just one kind of plant, while others enjoy a wide variety. Leafhoppers run sideways when disturbed and, as the name implies, can leap from plants; the adults also fly. Although populations can burgeon, leafhoppers aren't usually present in large enough numbers to cause great harm.

Target: Many plants.

Damage: Leaves are whitened and mottled; the edges may turn brown and curl upward, a condition called hopperburn. Severe infestations may cause stunting or leaf drop. Leafhoppers excrete honeydew, which attracts ants and fosters sooty mold (see page 104). The pests also spread diseases as they feed.

Life cycle: Adults lay yellowish, elongated, slightly curved eggs in the stems or leaf veins of host plants. The pests usually overwinter as adults in debris or weeds. There are several generations each year.

Control: Row covers, sanitation, weeding, yellow sticky traps, water jets, sulfur, soap sprays, oil sprays, green lacewings, pyrethrum, rotenone, sabadilla, malathion, diazinon, carbaryl, acephate.

Notes: Leafhoppers leave tiny, varnishlike spots of excrement on plants.

LEAF MINERS

This is a catchall name for certain moth, beetle, and fly larvae that tunnel between the upper and lower surfaces of leaves, ruining crops of leafy vegetables and disfiguring ornamental plants. Each species produces a characteristic pattern of winding trails or blotches, but specific identification of the guilty party isn't usually required for successful control. On vegetables, the most common leaf miners are the larvae of tiny black flies with yellow markings.

Target: Vegetables and many other plants.

Damage: Leaves are marred; infested seedlings may be stunted or killed.

Life cycle: Adults lay eggs just under the leaf surface. Larvae hatch and tunnel through the leaf, then drop to the soil or another protected place to pupate. There are usually several generations each year.

Control: Row covers, yellow sticky traps (adults); handpicking infested leaves, plastic mulch (larvae); tilling (pupae).

Notes: Plastic mulch prevents larvae from reaching the soil to pupate and exposes them to predators. If the controls listed above don't work, consult the local Cooperative Extension office for chemical controls for your area and crop.

LEAFROLLERS

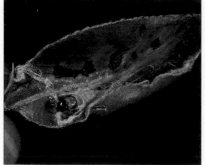

The name "leafroller" applies to the many species of caterpillars that roll leaves around themselves as they feed, creating a protective tube. Color and size vary with the species. Some leafrollers attack only one type of plant, while others eat numerous kinds. When disturbed, leafrollers wriggle backward and drop from the plant on a silken strand.

Target: Many plants.

Damage: Leaves are chewed on and tied together with webbing. Flowers and fruit may also be eaten.

Life cycle: Moths lay masses of eggs on plants; caterpillars hatch out, then feed and weave nests. The insects overwinter as eggs or pupae on plants. The number of generations a year varies, depending on the climate and the pest species.

Control: Trichogramma wasps, oil sprays (eggs); handpicking, spined soldier bugs, *Bt,* neem, diazinon, carbaryl, acephate (larvae).

Notes: Control the pest before it takes refuge in rolled leaves. If damage is light, pick off rolled leaves or squash them in place.

MEALYBUGS

Common on houseplants, these sap-feeding aphid relatives are also found outdoors in warm climates. The name "mealybug" refers to the female insects' powdery wax coating—a shield that prevents insecticides from penetrating. Colonies are often dense enough to make a cottony mound on leaves or, more typically, on stems.

Target: Soft tissues of most plants.

Damage: Leaves are distorted and yellowed; growth is stunted. Honeydew excreted by mealybugs attracts ants and promotes sooty mold (see page 104).

Life cycle: Each female lays about 600 eggs in a protective sac. Pale yellow nymphs hatch, then wander over the plant before settling down to feed; overlapping generations feed in the same cluster. Mealybugs can overwinter as nymphs or eggs.

Control: Handpicking, water jets, soap sprays, oil sprays, mealybug destroyers, green lacewings, pyrethrum, neem, malathion, diazinon, acephate.

Notes: Control ants (see page 72) or they'll protect mealybugs from predators. Hold off on persistent chemical sprays if you're counting on help from natural enemies.

MEXICAN BEAN BEETLES

Most common in the East, these copper-colored beetles, each with 16 black spots, resemble ladybird beetles in size and shape. The legless, ⅓-inch-long larvae are yellow, with six rows of long, black-tipped spines along their backs. Both adults and larvae feed on leaf undersides.

Target: Beans.

Damage: Leaves are chewed to lace; stems and pods may be eaten. Heavily infested plants may die.

Life cycle: In spring, adults feed for a short time before laying clusters of yellow eggs on leaf undersides. Larvae soon hatch; they feed, then pupate attached to leaves. Adults can overwinter in wooded areas or garden debris. There are one to four generations a year.

Control: Row covers, sanitation, handpicking, trap crop, rotenone, pyrethrum, malathion, carbaryl (adults); bean beetle parasites, spined soldier bugs, neem (larvae).

Notes: You can shake these beetles from their roosts onto a cloth, then discard them. Plant an early trap crop of beans, then release natural enemies.

MILLIPEDES

Although their name suggests otherwise, these nocturnal, hard-shelled, wormlike creatures don't have a thousand legs, but only up to a mere 400. Slow-moving and ranging from ½ to 2 inches in length, they're found in moist soil and garden debris; they often curl up when disturbed. A few types of millipedes are pests, but most benefit gardens by feeding on dead plant material. However, during periods of drought or when decaying matter is scarce, even the helpful species occasionally eat roots, tubers, or fruit resting on the ground. Some millipedes prey on soil-dwelling insects. If you think millipedes may be doing damage, check at night to make sure your suspicions are correct.

Target: Vegetables, lawns.

Damage: Roots are eaten, seedlings are severed, and holes are gnawed in tubers or fruit.

Life cycle: Millipedes lay clusters of translucent eggs in or on the soil. Adults can overwinter below ground. There is one generation a year.

Control: Keeping fruit off the ground, prompt harvesting, diazinon.

Notes: It isn't necessary to take action against millipedes unless they become numerous enough to do significant damage. Some millipedes emit a liquid that can irritate skin.

NEMATODES

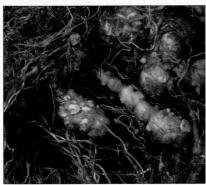

Members of a large family of roundworms too tiny to be seen, nematodes may be either friend or foe to the gardener. Some species—the parasitic type discussed on page 54—are beneficial organisms that attack soil-dwelling and certain other insects. But other nematodes are pests that slowly kill plants by feeding on the roots. Most produce root nodules (shown) that inhibit the uptake of nutrients; unlike the nitrogen-fixing nodules on legumes, these can't be flicked off with a thumbnail. Though nematodes can move only short distances under their own steam, they often spread through a garden by means of tools, transplants, water, and infested soil. Nematode damage is at its worst in warm, moist, sandy soils.

Target: Most plants, although many nematode species restrict their attentions to specific plant families.

Damage: Plants are yellowed, wilted, or stunted. Laboratory analysis of the soil is usually needed to confirm nematodes as the source of damage.

Life cycle: Each female lays a gelatinous egg mass; larvae develop in the eggs and begin feeding upon hatching. The life cycle takes about a month. Nematodes overwinter as eggs or larvae in the soil or on root nodules.

Control: Resistant varieties, soil solarization, crop rotation, marigolds, chitin, nitrogen fertilizer.

Notes: For reasons not fully understood, marigolds discourage nematodes. Previous recommendations called for solid stands of French dwarf marigolds (*Tagetes patula*) or African marigolds (*T. erecta*), but new evidence shows that interplanting marigolds with other crops also helps. Both chitin (see page 60) and quick-release nitrogen fertilizers encourage beneficial organisms that attack pest nematodes.

PARSLEYWORMS

This elusive, 2-inch-long, brilliantly colored caterpillar, which protrudes a pair of orange horns when disturbed, is rarely seen in gardens. Although technically a pest, it's so pretty you may want to overlook the paltry damage it does. Left alone, it will metamorphose into a gorgeous black-and-yellow swallowtail butterfly—another reason to admire it, not kill it.

Target: Carrots, celery, dill, parsley, parsnips, Queen Anne's lace.

Damage: Leaves and stems are nibbled.

Life cycle: Female swallowtails lay tiny, round green eggs on the leaf tips of host plants; small brown larvae hatch out, feed, and develop into large tricolored caterpillars. The insect overwinters as a pupa in cold climates, as an adult in warm climates. There are two to four generations a year.

Control: Handpicking.

Notes: Rather than killing the caterpillars, try moving them to a more expendable host plant.

POTATO TUBERWORMS

A pinkish white, ½-inch-long caterpillar with a brown head, this pest tunnels into the stems, tubers, and fruit of its target plants. It's a problem in the southern half of the United States.

Target: Primarily potatoes, but sometimes tomatoes, eggplant, and peppers.

Damage: Tunneling spoils tubers and fruit and causes shoots to wilt and die. If potato eyes turn pink with excrement, your crop is infested.

Life cycle: Small, narrow-winged, brown moths crawl through cracks in the soil to lay eggs on growing tubers; they also lay eggs on leaf undersides, debris, and soil. After feeding, the caterpillars pupate in debris or amid stored tubers. There are up to six generations a year.

Control: Crop rotation, mulching, destroying infested plants and tubers, sanitation.

Notes: Rotate tomato-family plants as a group. Eliminate cracks in the soil by applying a thick mulch or by hilling potato plants. Don't let newly harvested potatoes sit out on the ground overnight.

PSYLLIDS

The several species of these aphid-size insects, which suck sap primarily from leaf undersides, are sometimes called jumping plant lice. They feed on a variety of plants, including fruit trees; the pear psylla (shown) is the most damaging pear-tree pest in the United States.

Psyllids are extremely variable in appearance. The greenish or brownish adults could be mistaken for miniature cicadas; they have enlarged hind legs for leaping, clear wings, and long antennae. In the immature stage, some species are mobile and may be covered with a white, waxy coating, making them look like mealybugs. The young of other species resemble scale insects—they're immobile and feed in a single location.

Target: Many plants, including fruit trees and small fruits, acacia, boxwood, eucalyptus, eugenia, laurel, magnolia, pepper trees, potatoes, and tomatoes.

Damage: Leaves turn yellow, curl up, and eventually die; fruit is scarred. Liquid honeydew excreted by pear and blue gum psyllids attracts ants and encourages sooty mold (see page 104). Many types of psyllids transmit viruses.

Life cycle: The cycle varies with the species, but there are many overlapping generations each year. The pests overwinter as adults or eggs.

Control: Sulfur, diatomaceous earth, soap sprays, oil sprays, acephate.

Notes: Dust diatomaceous earth on foliage. Control ants (see page 72) or they'll protect psyllids from natural enemies.

ROOT MAGGOTS

These pests thrive in cool, moist, highly organic soils. The several species include cabbage and onion maggots, found primarily in the northern United States, and the seedcorn maggot, a widespread pest which destroys germinating seeds. The adults of most species look like gray houseflies. The white, legless, burrowing maggots have a pointy head and a blunt rear.

Target: Root vegetables, cole crops, tomato-family crops, seeds.

Damage: Tunneling ruins seeds, seedlings, and edible plant parts. Plants are stunted and prone to wilting. Some maggots spread diseases as they feed.

Life cycle: Adult flies lay white eggs at the base of host plants or in cracks in the soil. After hatching, the maggots tunnel into plant parts to feed, then form brown pupae in roots or soil. There are several generations a year; in warm climates, the pests may be active the year around.

Control: Crop rotation, row covers, tar paper collars (adults); destroying infested plants and roots, diatomaceous earth, diazinon (larvae); tilling (pupae).

Notes: Do your best to prevent these pests from establishing themselves, since there's no cure once an infestation begins. Reduce soil moisture and organic content. Mix diazinon into the soil before planting. Round or square tar paper collars, with a slit cut in the center to accommmodate the seedling, deter flies from laying eggs.

ROOT WEEVILS

Many species of root weevils cause trouble. The black vine weevil feasts on plants such as yew, rhododendrons, members of the rose family, and small fruits; the strawberry root weevil attacks a wide variety of plants, ranging from strawberries to arborvitae. The vegetable weevil (shown) feeds on carrots, tomatoes, spinach, and other plants.

Adults of all species are nocturnal, leaf-chewing, flightless beetles; they're about ⅓ inch long, with the typical long, curved weevil snout. When disturbed, they fall to the ground and play dead. The root-eating larvae are C-shaped, legless grubs with whitish or pinkish bodies and brown heads.

Target: Many plants.

Damage: Square or crescent-shaped notches are chewed in leaf edges (adults); plants wilt or die when roots are eaten (larvae).

Life cycle: The pests usually overwinter in the soil as grubs. After feeding heavily on roots, they pupate in late spring.

Adults soon emerge and feed at night for several weeks; then each female lays hundreds of eggs on host plants or on the soil. Grubs burrow into the ground and feed sparingly on roots throughout summer. The adults of many species feed on foliage until fall. There's usually one generation a year.

Control: Row covers, handpicking, sticky barriers, burlap trunk wraps, rotenone, pyrethrum, malathion, carbaryl, acephate (adults); destroying infested plants and roots, parasitic nematodes (larvae).

Notes: Gather burlap into 4-inch-wide accordion folds; then hold the material so the folds run horizontally and wrap it tightly around the base of any woody plant with leaves showing feeding notches. When the weevils crawl down, they'll nestle in the folds. Crush your catch daily. If you plan to use chemicals, it's helpful to identify the particular weevil, since not all products control every species.

ROSE CHAFERS

Unlike most insects named after particular plants, the rose chafer doesn't limit itself to one target. It's a general pest. The ½-inch-long adults—slender, long-legged, and light tan in color—feed in swarms, attacking flowering plants first, then moving on to other kinds of plants. The slim, white, ¾-inch-long grubs damage grass roots the same way Japanese beetle and June beetle grubs do (see page 82)—but only in sandy soils.

Target: Many flowers, including roses and peonies; some fruits and vegetables; lawns.

Damage: Leaves are chewed to lace and holes are eaten in flowers (adults); brown patches of lawn roll up (larvae).

Life cycle: After emerging in spring, adults feed for a few weeks, then lay eggs in sandy soil. The grubs hatch and feed near the surface, then tunnel deeper to overwinter. In spring, they move upward and feed for a time, then pupate near the surface. There is one generation a year.

Control: Row covers, handpicking, white sticky traps, rotenone, pyrethrum, carbaryl, acephate (adults); tilling, parasitic nematodes, diazinon (larvae).

Notes: Rose chafer grubs are not susceptible to milky spore disease.

SAWFLIES

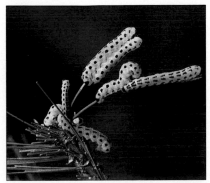

Unlike their bee and wasp relatives, sawflies are plant eaters, named for the way females use their egg-laying organs to saw slits in plants. Adult sawflies look like wasps, but they don't have a constricted waist, nor do they sting. They don't feed on plants—it's the larvae that do the damage. The various sawfly species include the cherry sawfly, pear slug, rose slug, and European pine sawfly; each pest attacks a narrow range of hosts. Larvae of the "slug" type do look like little slugs, complete with slime coating; other types resemble many-legged caterpillars.

Target: Many trees and shrubs.

Damage: Developing fruit is tunneled. Foliage is skeletonized—young leaves first, then older ones. Defoliated conifers may die.

Life cycle: Sawflies lay eggs in slits they cut in plants. After hatching, the larvae feed for several weeks, then pupate in the soil. The number of generations per year depends on the species.

Control: Soap sprays, oil sprays, rotenone, carbaryl, acephate (larvae); tilling (pupae).

Notes: Some species of sawflies—pear slugs, for example—seldom cause serious damage.

SCALE INSECTS

Varying in size, shape, and color, the many species of these aphid cousins look like bumps on bark, leaves, and fruit. All are equipped with an insecticide-resistant coat—a leathery or waxy material in the case of soft scales, a sturdier covering for armored scales. Young scales (called crawlers) move around the plant, but they settle down and become immobile as they mature.

Target: All plants.

Damage: The plant looks unsightly: it loses vigor and wilts, or new growth is distorted. Branches and even whole plants may die. Honeydew excreted by soft scales attracts ants and fosters the growth of sooty mold (see page 104).

Life cycle: Some scales lay eggs, often under a female's protective covering; other types bear live young. Armored scales usually have several generations a year, soft scales only one.

Control: Handpicking, green lacewings, mealybug destroyers, soap sprays, oil sprays, rotenone, malathion, diazinon, carbaryl, acephate.

Notes: Scrub scales off with a plastic scouring pad. Pesticides are most effective on the crawler stage. Since scales have many natural enemies, it's best to avoid broad-spectrum insecticides. Control ants (see page 72), which protect scales from natural enemies.

SLUGS & SNAILS

The many species of these mollusks, which vary in size and color, thrive in moist, shady surroundings. Snails have a shell and slugs don't, but both secrete a slimy mucus on which they glide along, leaving telltale silvery trails. The pests feed at night and on overcast days; when the sun shines, they go under cover. Slugs and some snails burrow into the soil, while other types of snails hide in debris or on plants. If conditions are too dry, a snail can seal itself in its shell and remain dormant for up to 4 years.

Target: All plants.

Damage: Large, ragged holes are eaten in leaves; seedlings are demolished.

Life cycle: Every slug and snail has both male and female sex organs, so any individual can lay clusters of tiny, gelatinous eggs in the soil. Some can even fertilize their own eggs. Depending on available food and moisture, a snail can take a few months or as a long as a few years to reach maturity. The pests can overwinter as eggs, but they're normally encountered as adults.

Control: Handpicking, trapping, sanitation, copper barriers, decollate snails, diatomaceous earth, snail bait (adults); tilling (eggs).

Notes: Overturned flowerpots, grapefruit halves, and boards laid on the ground make good traps. Instead of expensive beer traps, try the yeast traps described on page 45. Placing snail bait in stations keeps the poison dry and protects children, pets, and birds. (To make your own stations, cut holes in the sides of empty cans.) Salt is often suggested as a control, but it's bad for the soil, and the mollusks often recover.

SOD WEBWORMS

Sod webworms are the larvae of lawn moths, which hide in the day and fly in the evening and at night. When disturbed during daylight hours, the grayish white to tan moths make brief, erratic flights 1 to 2 feet above ground. They don't feed on lawns, but their slender, inch-long offspring chew on grass and, as the name "webworm" implies, spin silken tunnels in lawn thatch. Webworms are brown, gray, or green, with dark spots and long, stiff hairs.

Target: Lawns.

Damage: Grass stems are chewed off at the base, leaving brown patches in the lawn. Birds may make pencil-size holes in the turf while digging for the worms.

Life cycle: While in flight, moths drop eggs on the lawn. After hatching, the caterpillars feed by night on grass blades and stems; they can overwinter in the soil. There are usually two or three generations a year, but in the warmest areas, reproduction is continuous.

Control: Resistant varieties, dethatching, parasitic nematodes, *Bt*, soap sprays, pyrethrum, diazinon, carbaryl.

Notes: For a resistant lawn, plant ryegrasses or fescues infected with beneficial endophyte fungi (ask for these types at your nursery). To detect sod webworms, soak a square yard of lawn with a solution of 1 tablespoon household detergent in 1 gallon of water; if 15 or more worms appear, take action. Apply *Bt* about 2 weeks after you see moths flying over the lawn. Sod webworms have many natural enemies.

SOWBUGS & PILLBUGS

Oval and about ½ inch long, these seven-legged crustaceans are easy to tell apart: sowbugs have two tail-like appendages, and pillbugs curl into a tight ball if disturbed. Although adapted to life on dry land, the creatures thrive in damp conditions. To keep from drying out, they hide in dark, moist places during the day and feed at night. They're often wrongly blamed for plant damage, since they frequently spend the daylight hours in holes chewed by snails and other pests. Both sowbugs and pillbugs are largely beneficial: they feed primarily on decaying matter. It's true that they sometimes go after vegetation, but in general, they cause problems only when their numbers are high.

Target: Seedlings; ripe fruits and vegetables lying on damp ground.

Damage: Tender vegetation is chewed; holes are nibbled in ripe fruits and vegetables.

Life cycle: Each female lays eggs in a pouch on the underside of her body. The young, which look like smaller, paler versions of the adults, develop in the pouch. There are one to three generations a year.

Control: Keeping fruits and vegetables off the ground, prompt harvesting, diatomaceous earth, sanitation, diazinon, carbaryl.

Notes: To discourage sowbugs and pillbugs, water early in the day and use a coarse mulch that drains well. Instead of killing the pests, move them to a compost pile, where their work will be appreciated. If you see blue or purple individuals, leave them alone—they're infected with a virus and will eventually succumb.

SPIDER MITES

Too small to be seen clearly with the naked eye, these insect relatives flourish in hot, dry environments and on water-stressed plants. Unlike predatory mites (see page 55), spider mites suck plant juices and spin webs.

Target: All plants, but especially fruit trees and small fruits, cucurbits, tomatoes, and roses.

Damage: Leaves are stippled with yellow or fade to a bronze color. Plant parts may be distorted and covered with webs. Severely infested plants lose vigor and may die.

Life cycle: There are many generations a year; in warm climates, reproduction goes on nonstop the year around. The pests can overwinter as adults or as eggs on host plants.

Control: Water sprays, vacuuming, predatory mites, green lacewings, diatomaceous earth, sulfur, soap sprays, oil sprays.

Notes: To confirm the presence of mites, hold a sheet of white paper under a leaf and rap the stem sharply; any mites will tumble onto the paper and move around. Mist plants to create an unfavorable environment for mites; spray strong jets of water to knock the pests from their perches.

If you use sulfur, soap, or oil, you may need to spray every 7 to 10 days. Spider mites are resistant to most other pesticides—and if pesticides kill their natural enemies, the mites can increase uninterrupted, thanks to their very short life cycle.

SPITTLEBUGS

Frothy bubbles on a plant are a sure sign that these small, triangular, brownish or greenish creatures are in residence. The froth is actually a protective coating produced by the nymphs; the adults don't produce bubbles. Nymphs are more or less immobile, but adults are winged and will fly or hop away if disturbed. Spittlebugs do suck plant sap, but most species are harmless—though a few types can damage strawberries, holly, and pine trees.

Target: Many plants.

Damage: Usually no harm is done.

Life cycle: In spring, nymphs hatch from overwintering eggs and begin feeding on plants. There are one or two generations each year.

Control: Water jets, soap sprays.

Notes: Control isn't usually necessary.

SQUASH BUGS

Long-legged and mainly nocturnal, these true bugs are stubborn pests that favor cucurbits. Both adults and nymphs suck sap from stems and leaf undersides. The ½-inch-long, shield-shaped adults are yellowish to black, while the smaller, tear-shaped nymphs are yellowish green with red heads. Squash bugs stink when crushed.

Target: Primarily squash and pumpkins, but also cucumbers and melons.

Damage: Infested plant parts wilt and blacken; if the infestation is serious, the entire plant may die.

Life cycle: After overwintering in debris, adults lay clusters of shiny, reddish brown eggs on leaf undersides and stems. Nymphs that hatch from a single egg cluster feed together. There is one generation a year.

Control: Resistant varieties, sanitation, row covers, handpicking, vacuuming, trapping, soap sprays, rotenone, sabadilla, malathion, carbaryl.

Notes: Collect and destroy eggs. Put out boards or pieces of burlap as traps; destroy your catch each morning. Insecticides usually reduce populations only temporarily. Get rid of crop debris, since the pests will continue to feed on uprooted plants and clippings.

SQUASH VINE BORERS

An eastern pest, the squash vine borer is a 1-inch-long caterpillar with a white, accordionlike body and a brown head. It tunnels inside plant stems, cutting off the flow of nutrients and water. Infested vines may contain just one borer or as many as a hundred.

Target: Primarily squash, but occasionally cucumbers and melons.

Damage: Stems suddenly wilt and die. Greenish, sawdustlike excrement is piled around small entry holes near stem bases and on the ground.

Life cycle: When cucurbits begin to flower, red-and-black moths lay shiny, mahogany-brown eggs singly, usually on stems near plant bases. Caterpillars hatch in 1 to 2 weeks; they bore into stems and feed for about a month, then pupate in the soil. There are one or two generations a year. The pest overwinters underground as either a caterpillar or a pupa.

Control: Resistant varieties, timed plantings, row covers, foil collars around lower stem (adults); trichogramma wasps (eggs); trap crop, handpicking, sanitation, parasitic nematodes or *Bt* injected into stem, rotenone, carbaryl (larvae); tilling (pupae).

Notes: By planting early or late, you may manage to avoid peak midsummer damage. Chemical sprays, applied weekly when vines begin to run, kill caterpillars as they bore into stems. If only a few borers are infesting a stem, make lengthwise slits and pry the pests out; then heap soil to just above the highest slit so new roots will form. If stems are badly damaged, though, cut them off and destroy them, since borers may be inside. Increase irrigation to affected plants.

TARNISHED PLANT BUGS

This extremely agile insect owes its name to its mottled brown coloring. A true bug, it has a shield-shaped body with a triangular arrangement of white spots on the thorax. The long-legged, ¼-inch-long adults have a black-tipped yellow triangle at the end of each forewing; the nymphs are pale yellow. Both adults and nymphs suck sap from buds, fruits, and stems.

Target: Most fruits and vegetables.

Damage: Shoots are blackened; plant parts are deformed.

Life cycle: Nymphs hatch from elongated, curved eggs inserted into host plants. Adults as well as nymphs can overwinter in garden debris. There are three to five generations a year.

Control: Row covers, sanitation, white sticky traps, spined soldier bugs, soap sprays, sabadilla, malathion, carbaryl.

Notes: Getting rid of debris eliminates breeding spots. Tarnished plant bugs will migrate from recently cut hay or weeds to other areas, so be prepared to treat susceptible crops if your neighbors are planning to mow.

TENT CATERPILLARS

In some years, tent caterpillars seem to be everywhere; in others, they drop out of sight. Long-haired, wrinkly, and up to 3 inches long, these pests build gauzy nests in the forks of tree branches. Eastern types are black, with a white stripe along the back and blue dots on the sides; western species bear various markings, but all have white dashes. During the day, tent caterpillars dine outside their nests—unlike fall webworms, which expand their webs as they feed.

Target: Many trees and shrubs.

Damage: Leaves are chewed; branch forks are covered with webs. Plants may be defoliated.

Life cycle: Gray moths lay egg masses, which often encircle branches and twigs. Caterpillars hatch in early spring and begin making nests. They feed until early summer, then drop to the ground or lower themselves on silken threads, crawl to a protected spot, and pupate. There are up to four generations a year.

Control: Handpicking (eggs, larvae, pupae); trichogramma wasps (eggs); soap sprays, oil sprays, spined soldier bugs, *Bt*, diazinon, carbaryl, malathion, acephate (larvae).

Notes: When handpicking, wear gloves to avoid skin irritation. If you see caterpillars with shiny white "seeds" on their heads, leave them alone—the "seeds" are the eggs of parasitic tachinid flies. Spraying with *Bt* won't harm the flies, which can continue to live off the caterpillars' dead bodies. Before applying a pesticide, break up the tents with jets of water.

THRIPS

A few types of these narrow, barely visible insects are predators, but most are pests. Some kinds eat just about any plant, while others attack only a single species. Both the fringe-winged adults and the wingless nymphs scrape plant tissues and suck the juices. They often hide in buds and blossoms.

Target: Many ornamentals, especially flowers; most fruits and vegetables.

Damage: Plant parts are speckled, streaked, or distorted, but lack the webbing characteristic of mite damage. Black specks of excrement are visible. Some thrips transmit viruses.

Life cycle: Adults of most species lay eggs on plants, although some reproduce without mating. Because the entire life cycle takes just 2 to 3 weeks, there are many generations each year—and in warm climates, reproduction is continuous.

Control: Water jets, aluminum-foil mulch, yellow or blue sticky traps, green lacewings, predatory mites, sulfur, diatomaceous earth, soap sprays, oil sprays, pyrethrum, rotenone, ryania, neem, malathion, diazinon, acephate.

Notes: Dry plants are more likely to be attacked, so water adequately. Thrips hiding in plant parts are safe from contact pesticides.

TUSSOCK MOTHS

From one year to the next, tussock-moth caterpillars may go from troublesome to entirely absent—or vice versa. Both eastern and western species exist; all are colorful, 1¼-inch-long caterpillars with long bristles, spots along the length of the body, and a quartet of tufts.

Target: Many fruit, nut, and ornamental trees.

Damage: Leaves are chewed from the treetop down, sometimes to such an extent that the tree is defoliated. Fruit is scarred.

Life cycle: Wingless female moths emerge from cocoons in midsummer; they mate, lay hairy egg clusters on discarded cocoons in trees, and die. The eggs overwinter and hatch in spring. There are one to three generations each year.

Control: Handpicking, trichogramma wasps, oil sprays (eggs); *Bt*, carbaryl, acephate (larvae).

Notes: Search deciduous plants for the hairy, light brown cocoons on which eggs are laid. The pest population seems to peak every 7 to 10 years.

WHITEFLIES

Some 200 species of these pests cause problems. Like aphids (their close relatives), they're sap feeders, sucking plant juices from leaf undersides. The adults, which look like tiny white moths, fly up in a cloud when disturbed. Some of the nymphal stages resemble scale insects (see page 88). Whiteflies are found the year around in warm climates; in colder regions, you'll see them only during summer. Warm, still air is the perfect environment for whiteflies, making greenhouses a favored haunt—one species is even called the greenhouse whitefly.

Target: Many plants.

Damage: Heavily infested plants lose vigor and may turn yellow, but the biggest problem is usually honeydew, which attracts ants and fosters sooty mold (see page 104). Some adult whiteflies transmit viruses.

Life cycle: Adults lay minuscule eggs on leaf undersides. Yellowish nymphs hatch out and begin to feed; they lose their antennae and legs in the first molt, after which they're immobile throughout the remaining nymphal stages. All adults are winged, but only the females develop legs. There are many overlapping generations each year.

Control: Row covers, yellow sticky traps, water jets, handpicking, vacuuming, green lacewings, whitefly parasites, soap sprays, oil sprays, pyrethrum, neem, malathion, acephate.

Notes: Absolute control isn't needed, since plants can withstand moderate infestations. Use nitrogen fertilizers sparingly—the insects reproduce faster under high-nitrogen conditions. Pick off heavily infested leaves and destroy them. Whiteflies have many natural enemies, so it's best to forgo persistent chemicals. Control ants (see page 72) to keep them from protecting whiteflies from natural enemies.

WIREWORMS

The shiny, reddish brown larvae of the click beetle are the particular bane of gardeners who dote on root crops. Reaching a length of about 1½ inches, wireworms are hard-shelled and jointed, with three pairs of legs just behind the head. They're especially troublesome in gardens formerly planted as lawns. In hot, dry weather, they burrow deep into the soil.

Target: Many plants, especially root crops.

Damage: Underground plant parts are chewed; seeds may not germinate.

Life cycle: In spring or summer, adults lay eggs in the soil. Within a month, the larvae hatch and begin to feed. Some types mature in a year, while others take up to 6 years to pupate. There are overlapping generations. Both larvae and adults can overwinter in the soil.

Control: Soil solarization, trapping, tilling, diazinon.

Notes: Partially buried carrots or potato pieces serve as traps; remove and replace infested traps regularly. Till to a depth of at least 10 inches to kill the pests. Pesticides aren't usually needed.

Most gardeners appreciate wildlife—until the creatures start helping themselves to produce, digging up freshly planted bulbs, or otherwise threatening the garden. Fencing is the best way to deal with most four-legged marauders. Trapping (see page 46) and poisoning are practical in a few cases, but not always legal: though some animals (voles and pocket gophers, for example) are on their own, others are protected by law.

If you're unsure of a particular pest's status, check with your state game or conservation department.

Note that anticoagulant bait (a slow-acting poison that interferes with blood clotting) is recommended for controlling rodents. The word "anticoagulant" isn't always printed on the label—but if vitamin K_1 is listed as the antidote, the product is in fact an anticoagulant poison.

BIRDS

Only a few types, notably crows, blackbirds, house sparrows, and starlings, make pests of themselves at certain times. If you notice birds pecking at crops, take preventive measures before your feathered friends decide to designate the garden a permanent feeding station.

Target: Seeds, tender young plants, corn, fruit.

Damage: Ragged holes are pecked in leaves. Seeds and seedlings may disappear completely. Ripe fruit is partially eaten.

Control: Row covers, screens, bird netting, weeding.

Notes: Reflectors, fluttering objects, and scarecrows may reduce damage briefly, but birds soon learn to ignore such devices. For a more effective deterrent, protect seedlings with row covers or wire mesh screens, and drape bird netting over fruit trees and berry bushes. Use materials with approximately ½-inch-diameter holes and secure all edges. Suspend netting well away from plants; otherwise, birds will peck at fruit right through it.

To protect grapes, enclose each cluster in a paper bag (cut a few small holes in the bags for air circulation).

Observe pesky birds to see what habitats they prefer, then change or eliminate those areas. For example, if crows like to congregate on a brush pile, clear it away. It's helpful to eliminate weeds, since these plants' seeds attract some birds.

DEER

These graceful animals can wreck a garden in no time by nipping off blossoms and nibbling tender leaves and shoots. Fond of many flowering plants, deer also enjoy the foliage and fruit of just about anything you grow for your table. The types of plants most at risk will vary, depending on what the local deer like at the given moment (their tastes often change over time) and how hungry they are. As wild vegetation dries out, the animals spend more time foraging in gardens. They tend to establish trails, feeding mainly at dawn and dusk.

Target: Buds, leaves, and stems of many woody plants; most vegetables and fruits.

Damage: Chewed plant parts have jagged edges. Tree limbs may be gouged where males have used them for antler-polishing.

Control: Resistant plants, fencing, repellents.

Notes: Consult your Cooperative Extension office for a listing of plants unattractive to local deer. Fencing should be at least 8 feet high on level ground, even higher on a slope. As an alternative to a tall fence, try two parallel 4-foot-high fences spaced 5 feet apart. Wire mesh is the fencing material of choice, since deer can maneuver through 12-inch gaps between boards or wire strands. If fencing the entire garden isn't practical, you can set up barriers around individual plants or beds.

Almost any unfamiliar smell repels deer—but the novelty soon wears off, so change the type of repellent frequently. Bars of deodorant soap work fairly well. Hang one bar in each small tree, several in each large tree, so that no branches within the browsing zone (up to 6 feet high) are more than 3 feet away from a bar. Keep in mind that odor and taste repellents are often ineffective if deer are extremely hungry or your garden is particularly tasty.

DOGS & CATS

Pets can be a real annoyance in the garden. Cats like to dig in soft soil; undisciplined dogs romping through plantings are even more destructive.

Target: Garden soil.

Damage: Plants are trampled, soil is dug up, and urine and droppings are deposited in flower beds and on lawns (dogs); newly planted areas are used as a litter box (cats).

Control: Barriers, repellents.

Notes: Fences will keep out neighborhood dogs, but you'll still need to control your own pet. Training is the key: accustom your dog to other play areas, or use raised beds for flowers and vegetables and create paths for Fido to run along. Viny ground covers and thorny plantings deter many dogs.

To discourage cats, place wire mesh screens over newly planted beds or poke stakes into the bare patches between plants; you can remove the barriers as soon as foliage covers the ground. Some cats are partial to catnip, so you might try growing a patch well away from the plants you want to protect. Cats can clamber over most fences—but if you have one they can't squeeze under or through, you can make it fully catproof by attaching a chicken wire addition, cut side up, to the top. When a cat tries to climb over, it will fall back down again, since the wire won't support its weight.

Commercial odor repellents may work in rebuffing cats and dogs.

MOLES

Ridges formed in soft, moist earth and volcano-shaped mounds of finely pulverized soil with holes in the centers are sure signs of a mole invasion. These pests are insectivores, not rodents; they don't go after plants as food, but damage them while tunneling for insects and earthworms. About 5 to 7 inches long, moles have dark, velvety fur, a pointed snout, no visible ears or eyes, and outward-pointing claws for digging. Surface feeding burrows (the ridges you'll see) are used for short periods and may be abandoned soon after excavation; main runways are usually 8 to 18 inches below ground level. Most species are solitary. There's often an eruption of activity in summer, as young moles, driven out to fend for themselves, dig new tunnel systems.

Target: Soil-dwelling insects, earthworms.

Damage: Plants are heaved from the ground, roots are severed, and lawns are disfigured.

Control: Controlling grubs, barriers, trapping.

Notes: Eliminating grubs that feed on lawn roots may discourage moles; see listings for billbugs, Japanese beetles, June beetles, and rose chafers in "Insects & Their Relatives" (pages 72–92).

Protect plants from moles by lining planting holes and beds with ½-inch-diameter galvanized wire mesh. If your lawn suffers mole damage, try to salvage the ridged turf by pressing it back down right away, then watering thoroughly.

Trapping is the most effective way to get rid of moles. It's easiest to set a trap in a surface runway; to make sure the passage is still in use, poke small holes into one section, then see if the mole repairs the damage within a day. If you can't find an active surface tunnel, probe between two molehills to find a main runway (your probe will sink when you hit the right spot); then set your trap there.

Because tunnel systems are so extensive, flooding and gassing are rarely effective; poisoning is futile, too, since insectivores won't eat conventional poisons. Folk remedies include putting sticks of chewing gum in the tunnels, planting castor beans, and scattering dog hair, but there's no evidence that any of these methods works. If you see the ground surface heaving, you can try to nab the mole beneath with the two-shovel method: block the runway in front of your quarry with one shovel blade, then dig the animal out with the other shovel and dispatch it.

POCKET GOPHERS

If you see a plant wiggle, then disappear below ground, or if you notice a fan-shaped mound of soil blocked with a plug of earth, you know a pocket gopher is at work. Primarily found west of the Mississippi, these little rodents can ravage gardens. The animals are brown-furred and 6 to 12 inches long, with small eyes and ears, long claws, and teeth that grow up to 14 inches a year (gnawing keeps them worn down to a reasonable length). The "pocket" of the name refers to the fur-lined external cheek pouches used for carrying food.

Except when mating or bringing up young (offspring are driven out at 2 months of age), gophers live alone in tunnel systems occupying up to 2,000 square feet. Runways near the surface are for gathering food; deeper ones are for sleeping, storing food, and raising young. Lateral passages that rise to ground level at a 45-degree angle are disposal chutes for shunting excavated soil to the surface. A single garden may contain several intertwining tunnel systems. Your troubles aren't always over when you get rid of one pesky gopher, since another may move right into the original pest's tunnels.

Target: Roots, bulbs, tubers, seeds, and grasses.

Damage: Plants disappear or their roots are eaten. Young trees may be girdled and killed. Mounds cover plants and ruin lawns.

Control: Barriers, trapping, poisoning.

Notes: For fail-safe protection, line planting holes or beds with ½-inch-diameter galvanized wire mesh.

Trapping is the best way to get rid of gophers. Stand at a fresh mound with the plug in front of you; step back a foot and insert a probe. When you hit the main runway (it's usually 6 to 18 inches below ground), the probe will suddenly sink. Dig straight down and place two traps facing in opposite directions in the runway; anchor them to stakes for easy retrieval. For bait, use wads of grass, wild morning glory, or other green plant material. Some gardeners cover the hole above the traps, while others leave it open. If nothing happens within a day, there's probably no gopher in that tunnel; if you find a sprung trap with some fur in it, give up trying to trap that gopher (it's trap-wise).

If you plan to use poison, keep in mind that anticoagulant bait is safer than strychnine around children and pets. If you see new mounds more than 2 days after strychnine baiting or 10 days after anticoagulant baiting, you've struck out (but you can always try again). Gassing is seldom successful, since the animal usually escapes by walling off part of its burrow. Flooding, too, is generally ineffective. Other futile controls include chewing gum placed in the tunnels, noisemakers, and "gopher plants" (*Euphorbia lathyris*).

RABBITS

Throughout the year, gardens are subject to invasion by various species of hares (jackrabbits) and true rabbits (cottontails, brush rabbits, and related types). True rabbits, which give birth to blind, hairless young, usually dig their own shallow burrows or occupy those abandoned by other animals. Hares live aboveground: their offspring are born with fur coats and fully developed eyesight, and thus require less protection.

Target: Tender young growth.

Damage: Succulent leaves, shoots, and flowers are nibbled; young plants are often chewed to the ground. Bark on young trees and shrubs may be stripped, girdling the plants.

Control: Barriers, repellents.

Notes: The best way to deal with rabbit damage is to prevent it by fencing the offenders out. Erect a fence of ¾-inch-diameter wire mesh; make sure the mesh extends at least a foot below ground, and bend the bottom outward to keep rabbits from burrowing beneath. (Or add a buried wire mesh extension to existing fences.) The fence needn't be more than 2 feet high for rabbits, 3 feet high for hares; in snowy-winter climates, barriers should clear the snow line by these amounts. To protect individual plants, surround them with wire mesh cylinders, buried the requisite depth and braced so that rabbits can't push them in. Place row covers over seedlings. Commercial rabbit repellents may provide some short-term control. As for trapping, it usually doesn't do much good unless the whole neighborhood cooperates, since rabbits breed so prolifically. If you do want to trap, check first with your state game or conservation department—rabbits are considered game animals in most states.

RACCOONS

Dexterous of paw and hearty of appetite, this ring-tailed, masked marauder strikes at night. An excellent climber, it's omnivorous, dining on plants, insects, and snails. Raccoons are usually solitary except when mating and raising young.

Target: Many fruits and vegetables, especially corn and melons.

Damage: Corn stalks are bent and ears are munched on; melon flesh is scooped out. Lawns may be torn up.

Control: Sanitation, pruning, barriers, repellents, trapping.

Notes: To make your property less appealing to raccoons, harvest crops promptly, bring in pet food at night, and secure garbage cans tightly. To protect the vegetable garden, erect a 5- to 6-foot-high chicken-wire fence; then cover the top 2 feet with a band of sheet metal. A low, two-strand electric fence is also effective. To keep raccoons out of trees, prune lower limbs and wrap a 2-foot-wide metal band around the trunk, positioning the bottom of the band at least 2 feet off the ground. Some gardeners find that moth flakes repel raccoons. Playing a radio all night may also drive the pests away, but it won't earn you the goodwill of your neighbors. Trapping is the surest way to get rid of raccoons—but leave the job to a professional, since the animals can be vicious when captured.

SQUIRRELS

Both ground squirrels and tree squirrels can be troublesome in gardens. Ground squirrels are capable of climbing trees, but they live in underground burrows, where they store food, raise their young, and hide from predators. If a burrow is active, you'll see freshly excavated soil around the entrance (a hole about 4 inches wide). The animals wall themselves off when hibernating, but the plug isn't visible. Ground squirrels feed aboveground during the day, eating green plant material in spring during the breeding period, then switching to seeds and nuts later in the year.

Tree squirrels spend most of their lives in trees; they build nests in branch crotches or take up residence in convenient holes in tree trunks. Active all year, these agile animals can easily jump 6 feet. They live largely on seeds, nuts, fruits, and bark; they also gnaw into buildings and invade attics.

If you're not sure whether you're looking at a tree squirrel or a ground squirrel, just stamp your feet and see if the animal takes refuge in a tree or a burrow. Squirrels transmit diseases, so don't try to touch them or put your hand down a burrow.

Target: Fruits, seeds, nuts, grains, bark. Ground squirrels eat tender greens in spring.

Damage: Plant parts are eaten into; bark is gnawed.

Control: Prompt harvesting, barriers, trapping, poisoning, gassing.

Notes: Cover the soil around planting beds with sheets of 1- to 2-inch-diameter wire mesh; squirrels don't like to walk on it. Prune trees so the lowest branches are at least 6 feet above ground. Place 2-foot-wide metal bands around tree trunks, positioning them so the bottom of each band is at least 2 feet off the ground. Trapping is an effective control for ground squirrels, although you may need a permit to capture certain types. Gas cartridges or smoke bombs will kill squirrels in active burrows, but gassing is ineffective during hibernation, since the animals wall themselves off. Anticoagulant bait can also be used against ground squirrels, but check with authorities before trying to poison tree squirrels.

VOLES

Also known as meadow mice or meadow voles, these small, typically nocturnal rodents have gray to brown fur and strong, chisel-like teeth. Voles grow to about 7 inches long. They nest in shallow underground burrows or live aboveground in dense vegetation or leaf mulches; they sometimes take over abandoned mole tunnels. Narrow surface trails usually run between each animal's nest and its food sources. Though voles can climb, they rarely venture more than 2 feet off the ground, more typically feeding at or just below soil level. Vole populations rise and fall cyclically; when the numbers soar, the creatures are especially troublesome in gardens.

Target: Bulbs, tender vegetables and flowers, grasses, bark.

Damage: Plant parts are nibbled; seedlings may be eaten to the ground. Bark is gnawed or stripped.

Control: Weeding, sanitation, barriers, trapping, poisoning.

Notes: Eliminate favored vole habitats by weeding and pulling mulch and vegetation away from the base of trees and vines. Replace thick, fluffy mulches, which afford hiding places, with more compact materials. Fences of ¼-inch-diameter wire mesh, extending 6 to 10 inches below ground and at least 1 foot above the surface, will keep out voles. Wrap young tree trunks with mesh, burying the bottom edges to keep the animals from burrowing underneath. Also cover bulb beds with wire mesh, securing the edges. To trap voles, use ordinary mouse traps, placing them in burrows at right angles to the runway. Or poison the animals with anticoagulant bait set out in runways or next to burrows.

WOODCHUCKS

When these husky rodents cease hibernation and emerge from their 2- to 5-foot-deep burrows, spring has supposedly arrived. Found in the northern half of the United States, the animals are commonly called woodchucks or groundhogs in the East, marmots in the West. They're stocky and brown-coated, 16 to 20 inches long, with a furry tail and long, curved claws adapted for digging. As it excavates its burrow, a woodchuck creates a large telltale mound of soil.

Woodchucks are largely solitary (offspring are booted out of the burrow as soon as they can fend for themselves); they feed aboveground during the day instead of tunneling for food. Frequently, an abandoned burrow is quickly taken over by another woodchuck or other burrowing animal. Since woodchucks can transmit diseases, don't stick your hand into a burrow or try handling the animals.

Target: Tender vegetables, flowers, and succulent leaves; sometimes tree trunks.

Damage: Vegetation is chewed; woody stems may be gnawed.

Control: Barriers.

Notes: Deter woodchucks with a 3-foot-high wire mesh fence with the bottom foot bent outward and buried a few inches beneath the soil surface. Extend the mesh 1½ feet above the fence posts; that way, animals attempting to crawl over will simply be dumped back to the ground (the wire won't bear their weight).

PLANT DISEASES

The following pages include descriptions and control measures for over two dozen diseases. Some maladies are common everywhere; others you may never encounter unless you live in a particular region or grow certain plants. To decrease your chances of seeing any diseases at all, concentrate on cultivating a healthy garden—one where each plant receives the growing conditions it needs (see pages 21–39).

Because the status of many synthetic fungicides is uncertain today (see page 68), we have not listed these chemicals as possible controls for any of the diseases discussed here. For information on the products currently available to treat particular problems, consult your local Cooperative Extension agent; experienced nursery personnel can also help you decide how best to treat fungal diseases.

ANTHRACNOSE

The name "anthracnose" encompasses numerous diseases caused by fungi that flourish in wet weather; each fungus attacks only a narrow range of plants. Prevalent in the eastern and central United States, anthracnose diseases are characterized by cankers (sunken lesions) on leaves, fruit, and stems. Spore masses, which look like light pink slime, frequently ooze from the cankers. The spores are spread by wind and rain, as well as by gardeners handling wet plants. Anthracnose is also carried in infected seeds.

Target: Beans, cucurbits, and other vegetables; dogwood, ash, elm, maple, oak, sycamore, and other trees.

Damage: Cankers form on plant parts. Infected leaves often drop prematurely; diseased stems and twigs die back.

Control: Resistant varieties, crop rotation, certified disease-free seeds, sanitation, pruning, baking soda, copper fungicides, Bordeaux mixture.

Notes: Use certified disease-free seeds grown in the western United States. Avoid working among wet plants. Prune out infected branches. Apply baking soda (see page 59) or fungicide as soon as symptoms appear.

BACTERIAL WILT

Incurable and always fatal, this disease is caused by various soilborne bacteria that break down a plant's cells, producing debris that clogs the water-conducting vessels. Wilted plants may partially recover at night, making you think they simply need water—but watering won't help. The bacteria are often spread by infected seeds or transmitted by certain insects (cucumber beetles and flea beetles, for example) during feeding. As a definitive test for bacterial wilt, cut a stem in two; put the cut ends together and squeeze, then pull the pieces apart again. If a white, mucuslike thread forms, it's too late to save the plant. Sometimes you'll see the telltale gummy material oozing on its own from cracks in stems and leaves. Wilt-causing bacteria can survive in the soil for many years.

Target: Many vegetables (especially cucurbits, corn, and tomato-family crops), flowers, and other nonwoody plants.

Damage: Plants wilt and dry up—individual leaves first, then shoots and larger branches. Young stems or vines die quickly, older ones more slowly.

Control: Resistant varieties, sanitation, controlling the insects that spread the disease.

Notes: Remove and destroy diseased plants, then wash your hands thoroughly and disinfect all tools in a solution of nine parts water to one part bleach or rubbing alcohol. Control cucumber beetles and flea beetles (see pages 77 and 79).

BLACK SPOT

The characteristic black spots on infected leaves are patches of fungus, not dead tissue. The organism attacks only roses; it survives on canes and fallen leaves, then spreads easily during misty, foggy, or rainy weather. Black spot is common in regions where summer weather is warm and humid.

Target: Roses.

Damage: Roughly circular black spots with ragged yellowish borders appear on leaves. In severe cases, defoliation occurs.

Control: Resistant varieties, good air circulation, sanitation, baking soda, sulfur, lime sulfur.

Notes: Water early in the day; avoid working among wet plants. Remove and destroy infected leaves. Baking soda (see page 59) is particularly effective; apply it as soon as you notice symptoms, then repeat every 2 weeks or so.

BROWN ROT OF STONE FRUIT

Spread by wind and rain, this fungal disease enters the blossoms of fruit trees, then moves down the twigs. It survives on infected twigs and on "mummies" (shriveled fruit) to reinfect the tree the following year.

Target: Apricot, nectarine, peach, plum, and other stone fruits.

Damage: Blossoms wilt and decay; twigs crack and ooze sap. Fruit shows soft, brown spots, which may enlarge, rot, and become covered with spores.

Control: Bird netting, sanitation, pruning, prompt harvesting, copper fungicides, Bordeaux mixture, sulfur, lime sulfur.

Notes: Use netting to control birds, since wounds pecked in fruit can set the stage for infection. Destroy fallen fruit and leaves; pick rotten or shriveled fruit still on the tree. Prune to remove blighted twigs and to admit sunlight and improve air circulation. Spray with a fungicide during bloom; to prevent fruit rot, you can spray again 2 to 3 weeks before harvest.

CROWN GALL

Tumorlike growths near the soil line on a plant's roots or stem are telltale signs of this bacterial disease. Plants aren't always seriously harmed, but they often look horrid. The bacteria can survive in the soil or on dead tissue for 2 to 3 years.

Target: Rose-family plants, chrysanthemums, marigolds, euonymus, and many other plants.

Damage: Plants are usually weakened or stunted. Even if the galls don't damage the host plant directly, they may split apart and invite attack by other pests.

Control: Resistant varieties or treated plants; soil solarization.

Notes: Ask for plants treated with the microbial pesticide *Agrobacterium radiobacter*; or buy the pesticide from a garden supplier and apply it yourself. Avoid wounding plants when cultivating the soil around them.

DAMPING-OFF

Present in almost all soils, the various organisms referred to as damping-off fungi can kill seedlings—sometimes even before they emerge—and rot potato seed pieces. Healthy seedlings resist infection, though, and plants generally become less vulnerable with age. Overwatering and poor air circulation foster the growth of damping-off fungi, and plants are more susceptible to the disease in wet, poorly drained, cold soils with a high nitrogen content.

Target: Seeds, young plants.

Damage: Seedlings fail to sprout; or they may die soon after emerging.

Control: Soil solarization, sanitation, fresh seeds, treated seeds, improving drainage, proper planting and watering.

Notes: The fresher the seeds, the less likely they are to fall victim to the disease. You may also want to try seeds treated with a fungicide. Plant seeds when the soil temperature favors growth; set them at the right depth, leaving adequate space between seeds. Don't overwater, and hold off on fertilizing until seedlings have true leaves (not just initial seed leaves). To improve your chances of success, you can start seeds indoors in a sterile planting mix, then transplant the seedlings outdoors. Sanitize your propagation tools, pots, and flats in a solution of nine parts water to one part bleach or rubbing alcohol.

DOWNY MILDEW

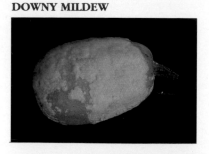

Each of the many fungi causing downy mildew attacks a narrow range of hosts. Spread by wind, rain, and infected seeds, these fungi all require wet conditions, and most must also have cool temperatures (the exception is downy mildew of cucurbits, which can germinate at temperatures as high as 90°F).

Target: Many plants, including snapdragons, pansies, roses, marigolds, strawberries, grapes, and most vegetables.

Damage: Leaves show gray, white, or purplish fuzz on their undersides, yellow blotches on the top surfaces.

Control: Resistant varieties, crop rotation, sanitation, good air circulation, pruning, copper fungicides, Bordeaux mixture, sulfur.

Notes: Keep foliage as dry as possible.

DUTCH ELM DISEASE

The tall, stately elms which once graced American streets have been felled by this devastating fungus disease. Its origins are unclear; the "Dutch" in the name refers to the nationality of the botanist who discovered the fungus. For years confined to the East and Midwest, the disease spread slowly across the country, reaching the West in the 1970s.

Dutch elm disease kills by clogging a tree's water-conducting tissue. It's transmitted by the elm bark beetle, and can also move from infected to nearby healthy trees when the roots rub against each other. Chinese, Siberian, and a few other Asian elm species are not susceptible, and plant breeders are working to develop new resistant varieties with the majestic shape of the American elm.

Target: American and European elms.

Damage: The leaves of one or several branches wilt, turn yellow, and drop; then the wood, and eventually the entire tree, dies as well.

Control: Resistant varieties, controlling bark beetles.

Notes: Maintaining a healthy tree is the best way to discourage bark beetles (see page 74). Once a tree is infected, there's no cure. Remove and destroy diseased wood to keep the fungus from spreading. Certain fungicides can be injected into healthy susceptible elms as a preventive; consult a tree care specialist.

FIREBLIGHT

A bacterial disease, fireblight affects only members of the rose family. The infection enters through the blossoms, then spreads with the help of pollenizing insects and splashing water from rain or sprinklers. Fireblight is favored by temperatures above 60°F and high humidity caused by rain, dew, fog, or irrigation; in fact, if these conditions persist for at least 48 hours during bloom, infection is a near certainty. The bacteria survive in blighted twigs and cankers.

Target: Pear and quince are most susceptible; apple, crabapple, and pyracantha are frequently damaged; hawthorn, spirea, cotoneaster, toyon, serviceberry, loquat, and mountain ash are occasional victims.

Damage: Leaves, shoots, and developing fruit wilt and blacken as if scorched by fire. Dark, sunken cankers may form in large branches.

Control: Resistant varieties, pruning, copper fungicides, Bordeaux mixture, streptomycin.

Notes: Prune off diseased growth, making cuts at least 6 inches below the infection on the smaller branches, at least 12 inches below blighted areas on the larger limbs. After each cut, disinfect pruners in a solution of nine parts water to one part bleach or rubbing alcohol. Apply a fungicide or streptomycin when about 10 percent of the flowers are in bloom; repeat at 4- or 5-day intervals until bloom ends. Copper can cause russeting in pears.

FUSARIUM WILT

Most active in warm soils, this soilborne fungal disease plugs a plant's water-conducting vessels. Each of the many fusarium strains attacks a narrow range of hosts; the afflicted plants are primarily herbaceous rather than woody. Spores can survive in the soil for about 10 years.

Target: Many plants, although each strain is specific to only a few hosts.

Damage: Leaves and stems wilt, turn yellow, and eventually die. A plant may initially show symptoms on just one side or one branch.

Control: Resistant varieties, soil solarization, crop rotation, timed plantings, sanitation.

Notes: Because the spores can survive so long in the soil, a very lengthy rotation is needed. Plant peas and other cool-season crops as early as possible, so the plants can mature before the fungi become active. Dig up and destroy infected plants.

GRAY MOLD

The gray mold that afflicts garden plants is the very same organism you'll often see on elderly boxed strawberries at the supermarket. Also known as botrytis blight, the disease is caused by a fungus that thrives in shady conditions and crowded plantings; it's especially common in cool, humid climates. The spores are spread by wind and water. Gray mold usually begins on plant debris and old plant tissue (such as spent flowers and overripe fruit), then invades actively growing tissue.

Target: Many flowers, fruits, and vegetables.

Damage: Soft, tan to brown blotches form on plant parts, then become covered with a coarse, grayish mold. The mold may turn slimy as the tissue beneath it rots.

Control: Good air circulation, sanitation, sulfur.

Notes: Water early in the day, so plants have time to dry by nightfall. Remove dead leaves and flowers to eliminate possible breeding grounds for infection; pick off diseased tissue. Ask your Cooperative Extension agent or an experienced nurseryman about currently available fungicides effective against gray mold.

MOSAIC VIRUSES

There's a multitude of these common viruses; each kind affects a single host or plant family. Many are spread by insects. Like all viruses, mosaics are incurable—although sometimes they simply create interestingly patterned leaves without significantly reducing a plant's vigor.

Target: Many plants.

Damage: Leaves are mottled or streaked. Other possible symptoms include distorted or stunted growth, reduced yield, and poor fruit quality.

Control: Resistant varieties, certified disease-free stock, crop rotation, weeding, destroying infected plants.

Notes: Resistant varieties may be available for certain crops. Control weeds, since many serve as reservoirs for viruses. To avoid spreading tobacco mosaic virus, don't smoke around tomato-family plants and wash your hands before touching them. Controlling aphids and other virus-spreading insects is a good idea in theory, but total elimination of these pests isn't a practical goal.

LAWN DISEASES

If you focus on good care (see pages 32–33), you may be able to avoid lawn diseases entirely: healthy, vigorous lawns are much less vulnerable to infection than poorly grown ones. If you're planting a new lawn, ask about disease-resistant grasses appropriate for your area. Also check on the fungicides currently available for treating specific diseases. The following disorders commonly affect lawns; see also powdery mildew, rust, and southern blight (all discussed on page 104). Insects, too, can cause lawn damage; see "Lawns" on page 110.

Fairy rings. Rings of dark green grass, often bordered with small tan mushrooms, are clear signs of this condition. Areas of dead growth may appear inside the rings. Fairy rings thrive in mild, moist weather. Because the fungi grow on organic matter in the soil, they don't harm the grass directly, but they may restrict water flow. To correct the problem, soak the rings with water every day for a month, remove thatch and aerate the lawn, and fertilize. By greening up the rest of the grass, you'll obscure the rings.

Leaf spot. This condition causes patches of thin, brown grass; the blades have small oval spots with bright red centers and darker borders. Most common on bluegrass and ryegrass, leaf spot is caused by various fungi (some formerly known as helminthosporia leaf spot fungi). Use a fungicide when you

FAIRY RINGS

LEAF SPOT

PYTHIUM BLIGHT

OAK ROOT FUNGUS

Also called armillaria, shoestring, or mushroom root rot, this disease is widespread on the West Coast and in the Southeast. Especially active in wet soils, the fungus invades the roots of many plants, especially those that are weakened or stressed. It can live on native oaks without doing any damage—as long as the trees aren't watered during the dry season. Since symptoms of the disease vary, just be on the lookout for the actual fungus: creamy white, fan-shaped layers, often with a strong mushroom odor, growing beneath the bark near or below the soil level. Clusters of honey-colored mushrooms may appear around the base of the tree in fall or winter.

Oak root fungus doesn't spread through the soil; it moves from one tree to another when healthy roots come into contact with infected ones. It can survive for at least 30 years in dead roots below ground.

Target: Primarily woody plants, especially oaks and fruit trees.

Damage: The trunk is girdled. Dull, yellowed, or wilted foliage is usually the first sign of trouble; infected trees usually die back slowly.

Control: Resistant plants.

Notes: There is no cure. Remove dead or badly stricken plants, trying to dig out as much of the roots as possible. You may be able to prolong the life of a plant by uncovering the infected root crown, then leaving it exposed to air and giving little or no water.

first notice damage. Too-lush lawns and those under stress from excessive water or fertilizer, short mowing, and thatch are most susceptible.

Pythium blight. Also known as cottony blight and grease spot, this fungal disease is a serious problem for bluegrass and ryegrass; top-seeded lawns and those sown too thickly are especially vulnerable. The disease shows up as irregular areas of dark green, water-soaked grass; the blades are sometimes streaked with white fungal threads. Affected spots spread quickly; as the lawn dies, the fungal threads disappear and the grass turns red-brown, then tan. Correct the problem by aerating the lawn, improving drainage, and, if possible, reducing the amount of shade. Fungicide treatment is another option.

Slime molds. Unsightly but harmless, this type of fungal growth is found in the East and on the Pacific Coast; it flourishes in humid conditions on all types of lawns. After irrigation or a heavy rain, affected grass blades are covered with tiny, powdery balls that may be bluish gray, black, or yellow. The fungi feed on decaying matter in the soil; they damage grass by blocking light, causing infected areas to yellow. To correct the problem, hose or sweep grass clean.

Snow molds. Snow molds show up in at least two forms; both prosper in wet, cold weather, especially if grass remains soaked for long periods. Gray snow mold causes irregular dead, bleached patches up to 2 feet across; gray mold is visible on the grass. Pink snow mold produces circular, light brown patches, sometimes blotched with pink fungus. Affected grass pulls up easily. To correct the problem, aerate the lawn, improve drainage, avoid high-nitrogen fertilizer late in fall, and reduce snow pileup.

Tip burn. Also called septoria leaf spot, this fungal disease appears in cool, wet weather in spring and fall. Bermuda and most cool-season grasses are susceptible. Blade tips turn pale yellow to gray, with tiny black dots and red or yellow margins. Mowing removes diseased tips; fungicide is an option, too.

SLIME MOLD

SNOW MOLD

TIP BURN

OAK WILT

This deadly fungal disease, which blocks a tree's water-conducting vessels, has killed millions of oaks east of the Rockies, especially in Texas and the Great Lakes states. The fungus is spread by oak bark beetles and by contact between healthy and infected roots.

Target: Oaks, especially red oaks.

Damage: Starting from the treetop down, leaves wilt, turn dull, curl, dry, and drop. The tree eventually dies.

Control: Resistant varieties, controlling bark beetles.

Notes: There is no cure. Remove and destroy infected trees, including stumps. Avoid pruning in spring, when bark beetles are most active. If a tree is wounded, paint over the wound immediately to deter beetles from entering. Certain fungicides can be injected into healthy trees as a preventive; consult a tree care specialist.

PEACH LEAF CURL

Most active during cool, rainy springs, this fungal disease slowly weakens infected trees if left unchecked. Afflicted leaves show a coating of white spores; these can reinfect the host tree the following year.

Target: Peach, nectarine.

Damage: Midribs thicken, causing the leaves to pucker and curl; leaves are tinged with red or yellow and drop early. Fruiting is poor; fruit may be spotted.

Repeated infections cause branches to die back.

Control: Sanitation, copper fungicides, Bordeaux mixture, lime sulfur.

Notes: Apply fungicide in late winter, before new growth appears. You can also control infection by keeping trees covered with plastic film until the leaves unfurl (this is easiest to do with genetic dwarf varieties). Pick off and destroy infected leaves.

POWDERY MILDEW

While most fungi thrive in moist surroundings, the organisms causing powdery mildew flourish in dry conditions; they also prefer warm days, cool nights, and shade. The various members of the group attack different hosts and survive from year to year on perennial plants.

Target: Many plants, especially cucurbits, roses, phlox, dahlias, beans, peas, grapes, small fruits and fruit trees, euonymus, and bluegrass lawns.

Damage: Gray or white circular patches appear on leaves; then whole leaves become powdery white and distorted.

Infected leaves may drop, and the entire plant may be weakened or killed. The flavor of peas, squash, and melons may be affected.

Control: Resistant varieties, good air circulation, sanitation, pruning, misting, baking soda, copper fungicides, Bordeaux mixture, sulfur, lime sulfur.

Notes: Since dry air favors the disease, mist susceptible plants. Spray with baking soda (see page 59) or fungicide when symptoms first appear. Prune off and destroy infected tissue.

RUST

Most of the fungi known as "rust" attack ornamentals. Each is specific to a certain type of plant, although some—cedar-apple rust, for example—require two hosts. The spores are spread by wind and splashing water.

Target: Many plants, including roses, hollyhocks, and snapdragons; lawns.

Damage: Rust first appears as pustules on the undersides of leaves; the lesions are usually orange-yellow in color, but may also be brown or purple. Eventually, the upper leaf surfaces become mottled with yellow. Severely infected plants are stunted and may die.

Control: Resistant varieties, sanitation, good air circulation, Bordeaux mixture, sulfur, lime sulfur.

Notes: Clean up fallen debris; remove rust-infected leaves remaining on plants. Water overhead only in the morning on sunny days.

SOOTY MOLD

As you might guess from the name, this mold forms a black coating on the leaves of afflicted plants. The organism lives on a plant's natural secretions and on honeydew, the plant sap excreted by aphids, leafhoppers, mealybugs, psyllids, scale insects, and whiteflies. Cool, moist conditions encourage sooty mold.

Target: All plants.

Damage: Sooty mold harms plants by blocking light to the leaves, thus hindering photosynthesis.

Control: Controlling honeydew-producing insects, cleaning the plant.

Notes: For information on controlling the insects mentioned at left, consult the listings in "Insects & Their Relatives" (pages 72–92). You can wipe or hose the mold from plants; rain will also remove it.

SOUTHERN BLIGHT

A soilborne fungal disease that rots plant stems, southern blight thrives in the warm soils and wet weather of the South. It's also known as southern wilt, sclerotium root rot, and mustard-seed fungus; the last name refers to the organism's small, yellowish resting bodies. The fungus can survive for many years without a host.

Target: Many flowers and vegetables (especially peanuts and tomato-family plants); lawns.

Damage: White, cottony growth appears on plant stems near soil level and often spreads to the surrounding soil. As the flow of water through its stems grows progressively more restricted, the plant wilts, yellows, and ultimately dies.

Control: Crop rotation, soil solarization, enriching the soil, aluminum-foil collars, removing infected plants and soil.

Notes: Plan a 2- to 3-year crop rotation, alternating susceptible plants with corn or other immune plants. Since nitrogen deficiency promotes the fungus, add organic matter to the soil and increase its nitrogen content. Protect plants by wrapping the main stems with foil from just above the roots to about 2 inches above the soil. Dig up and destroy infected plants, including at least 8 inches of soil on all sides.

TEXAS ROOT ROT

Also called cotton root rot, this fungal disease is common in the warm, alkaline soils of the Southwest. It destroys the outer portions of roots, cutting off a plant's water supply. The fungus can survive in the soil for about 5 years.

Target: More than 2,000 plant species. Grasses and other monocots—plants with one seed leaf instead of two—are not affected.

Damage: Typically, damage isn't noticed until it's too late to save the plant: the first symptom—sudden wilting of leaves in summer, without leaf drop—signals that at least half the root system is affected. Rotted roots are covered with yellowish fungal growth.

Control: Resistant plants, acidifying the soil.

Notes: You may be able to salvage an infected tree or shrub by pruning off half the foliage, then loosening the soil to the drip line and covering it with 2 inches of composted manure. On top of the manure, scatter ammonium sulfate at a rate of 1 pound per 10 square feet; also apply soil sulfur at the same rate. Form a watering basin and soak the soil 3 to 4 feet deep. In the future, increase the organic content of planting beds and add soil sulfur to decrease alkalinity.

VERTICILLIUM WILT

A widespread fungus, verticillium is a soilborne organism that plugs water-conducting vessels. Though it thrives in cool, moist soils, it usually doesn't reveal its presence until the weather turns warm and dry.

Target: More than 200 plant species, including tomatoes, potatoes, strawberries, and Japanese maple.

Damage: One branch or one side of a plant typically wilts. The leaves yellow, starting from the edges and progressing inward; then they turn brown and die. The wilt progresses upward or outward from the base of the plant or branch; the tissue inside dead branches is discolored. Often, the entire plant dies.

Control: Resistant plants, soil solarization, new soil, reducing nitrogen fertilizer, pruning.

Notes: Don't till before solarizing: the fungus is usually in the top 6 inches of soil, and that's about how far the effects of solarization go. As an alternative to solarization, use clean, new soil in raised beds or containers. Crop rotation is ineffective, since the fungus can survive in the soil for 20 years. Excess nitrogen favors verticillium wilt, so cut back on nitrogen use. Pruning infected limbs may help a stricken tree recover. Maintain adequate soil moisture.

WATER MOLD ROOT ROT

Members of this group of fungi (various species of *Phythophthora* and *Pythium*) thrive when water stands too long around plant roots. The rot invades the roots and moves up the stems, sometimes girdling them at or below ground level. A dark discoloration is usually visible between healthy and infected tissue. The disease is very common in irrigated areas of California and the arid southwest.

Target: Many plants.

Damage: Leaf color usually dulls; leaves may yellow and wilt. The plant loses vigor and sometimes dies, either quickly or over the course of months.

Control: Soil solarization, improving drainage, proper planting and watering.

Notes: There is no cure, so concentrate on prevention. Plant at the right depth in well-drained soil; provide only enough moisture for healthy growth.

SOME COMMON PLANTINGS & THEIR PESTS

If you know which pests are likely to attack certain plants, you can get a head start in identifying garden problems. The entries below list the major troublemakers in strict alphabetical order, not according to the seriousness of the damage they cause. To narrow down the field of possible culprits, refer back to the photographs and descriptions in the earlier sections of this chapter.

SEEDLINGS	■ Armyworms, cutworms, earwigs, flea beetles, root maggots, slugs and snails, sowbugs and pillbugs ■ Birds, cats, deer, rabbits, squirrels, voles, woodchucks ■ Damping-off

VEGETABLES

Asparagus	■ Aphids, asparagus beetles, harlequin bugs, Japanese beetles, slugs and snails, spider mites, tarnished plant bugs ■ Deer ■ Gray mold, fusarium wilt, rust
Bean	■ Aphids, armyworms, blister beetles, corn earworms, cucumber beetles, cutworms, European corn borers, flea beetles, harlequin bugs, Japanese beetles, June beetles, leafhoppers, leaf miners, mealybugs, Mexican bean beetles, nematodes, slugs and snails, spider mites, tarnished plant bugs, thrips, whiteflies, wireworms ■ Deer, rabbits, voles, woodchucks ■ Anthracnose, damping-off, downy mildew, fusarium wilt, gray mold, mosaic viruses, powdery mildew, rust, southern blight
Carrot	■ Blister beetles, flea beetles, leafhoppers, nematodes, parsleyworms, root weevils, thrips, wireworms ■ Deer, pocket gophers, rabbits, woodchucks ■ Mosaic viruses, powdery mildew, southern blight
Celery	■ Aphids, cabbage loopers, flea beetles, leafhoppers, nematodes, parsleyworms, root weevils, slugs and snails, spider mites, tarnished plant bugs, thrips ■ Deer, rabbits, woodchucks ■ Fusarium wilt, mosaic viruses
Cole crops	■ Aphids, armyworms, blister beetles, cabbage loopers, cutworms, earwigs, flea beetles, harlequin bugs, imported cabbageworms, leaf miners, nematodes, root maggots, root weevils, slugs and snails, tarnished plant bugs, thrips ■ Deer, rabbits, voles, woodchucks ■ Downy mildew, fusarium wilt, mosaic viruses, powdery mildew
Corn	■ Aphids, armyworms, blister beetles, chinch bugs, corn earworms, cucumber beetles, cutworms, earwigs, European corn borers, flea beetles, harlequin bugs, Japanese beetles, June beetles, nematodes, wireworms ■ Birds, deer, raccoons, squirrels ■ Bacterial wilt, rust
Cucurbits (cucumber, melon, pumpkin, squash)	■ Aphids, blister beetles, cucumber beetles, flea beetles, harlequin bugs, leafhoppers, nematodes, spider mites, squash bugs, squash vine borers, tarnished plant bugs, thrips, whiteflies ■ Birds, deer, raccoons, squirrels, voles, woodchucks ■ Anthracnose, bacterial wilt, damping-off, downy mildew, mosaic viruses, powdery mildew, verticillium wilt
Eggplant	■ Aphids, blister beetles, Colorado potato beetles, flea beetles, harlequin bugs, hornworms, leafhoppers, nematodes, potato tuberworms, spider mites, whiteflies ■ Anthracnose, bacterial wilt, mosaic viruses, powdery mildew, southern blight, verticillium wilt
Lettuce	■ Aphids, armyworms, cabbage loopers, corn earworms, flea beetles, harlequin bugs, leafhoppers, leaf miners, nematodes, slugs and snails, tarnished plant bugs, whiteflies ■ Birds, deer, pocket gophers, rabbits, woodchucks ■ Gray mold, powdery mildew, mosaic viruses, southern blight

Onion	■ Blister beetles, June beetles, nematodes, root maggots, thrips, wireworms ■ Pocket gophers, squirrels, voles ■ Downy mildew, fusarium wilt, rust, southern blight
Pea	■ Aphids, armyworms, blister beetles, cabbage loopers, corn earworms, cucumber beetles, flea beetles, leaf miners, nematodes, spider mites, thrips ■ Birds, deer, rabbits, voles, woodchucks ■ Anthracnose, downy mildew, fusarium wilt, mosaic viruses, powdery mildew
Peanut	■ Armyworms, cucumber beetles, flea beetles, leafhoppers, nematodes, spider mites ■ Deer, squirrels, voles ■ Southern blight, verticillium wilt
Pepper	■ Aphids, blister beetles, Colorado potato beetles, corn earworms, European corn borers, flea beetles, hornworms, leafhoppers, leaf miners, leafrollers, nematodes, potato tuberworms, spider mites, whiteflies ■ Anthracnose, bacterial wilt, fusarium wilt, mosaic viruses, southern blight
Potato	■ Aphids, blister beetles, cabbage loopers, Colorado potato beetles, cucumber beetles, European corn borers, flea beetles, harlequin bugs, hornworms, June beetles, leafhoppers, leaf miners, nematodes, potato tuberworms, psyllids, root maggots, root weevils, tarnished plant bugs, wireworms ■ Pocket gophers, voles ■ Anthracnose, bacterial wilt, mosaic viruses, southern blight, verticillium wilt
Spinach	■ Aphids, blister beetles, cabbage loopers, flea beetles, imported cabbageworms, leafhoppers, leaf miners, nematodes, root weevils ■ Birds, deer, rabbits, squirrels, woodchucks ■ Downy mildew, fusarium wilt, mosaic viruses, rust, verticillium wilt
Tomato	■ Aphids, armyworms, blister beetles, cabbage loopers, Colorado potato beetles, corn earworms, cucumber beetles, cutworms, European corn borers, flea beetles, hornworms, leafhoppers, nematodes, potato tuberworms, psyllids, slugs and snails, spider mites, thrips, whiteflies ■ Squirrels, voles ■ Anthracnose, bacterial wilt, fusarium wilt, gray mold, mosaic viruses, powdery mildew, southern blight, verticillium wilt
FRUITS	
Apple	■ Aphids, borers, cankerworms, cicadas, codling moths, cucumber beetles, curculios, fall webworms, fruit flies, gypsy moths, Japanese beetles, June beetles, leafhoppers, leaf miners, leafrollers, mealybugs, psyllids, sawflies, scale insects, spider mites, tarnished plant bugs, tent caterpillars, thrips, tussock moths ■ Birds, deer, pocket gophers, rabbits, squirrels, voles ■ Crown gall, fireblight, mosaic viruses, oak root fungus, powdery mildew, rust, southern blight
Blackberry & raspberry	■ Aphids, borers, cutworms, fruit flies, Japanese beetles, leafhoppers, leafrollers, psyllids, root weevils, sawflies, scale insects, spider mites, thrips, whiteflies ■ Birds, deer, rabbits, squirrels ■ Anthracnose, crown gall, downy mildew, gray mold, oak root fungus, powdery mildew, rust, verticillium wilt
Blueberry	■ Curculios, fruit flies, nematodes, scale insects ■ Birds, deer, squirrels ■ Gray mold, powdery mildew
Cherry	■ Aphids, borers, cankerworms, curculios, fall webworms, fruit flies, gypsy moths, Japanese beetles, leafrollers, nematodes, sawflies, scale insects, spider mites, tent caterpillars, tussock moths ■ Birds, deer, pocket gophers, voles ■ Brown rot of stone fruit, crown gall, gray mold, oak root fungus, powdery mildew, verticillium wilt

Grape	■ Aphids, armyworms, borers, Japanese beetles, leafhoppers, leafrollers, mealybugs, nematodes, rose chafers, scale insects, spider mites, thrips, whiteflies ■ Birds, deer, pocket gophers, rabbits, voles ■ Crown gall, downy mildew, gray mold, mosaic viruses, oak root fungus, powdery mildew
Peach & nectarine	■ Aphids, borers, cankerworms, cicadas, curculios, fall webworms, fruit flies, Japanese beetles, June beetles, leafrollers, mealybugs, nematodes, scale insects, spider mites, tarnished plant bugs, tent caterpillars, thrips, tussock moths ■ Birds, deer, pocket gophers, voles ■ Brown rot of stone fruit, crown gall, mosaic viruses, oak root fungus, peach leaf curl, powdery mildew, verticillium wilt
Pear	■ Aphids, borers, cankerworms, codling moths, curculios, flea beetles, fruit flies, leaf miners, leafrollers, mealybugs, psyllids, sawflies, scale insects, spider mites, tarnished plant bugs, tent caterpillars, thrips, tussock moths ■ Birds, deer, pocket gophers, rabbits, voles ■ Crown gall, fireblight, oak root fungus, rust
Plum & prune	■ Aphids, borers, cankerworms, curculios, fruit flies, Japanese beetles, leafrollers, mealybugs, sawflies, scale insects, spider mites, tent caterpillars, tussock moths ■ Birds, pocket gophers, rabbits, voles ■ Brown rot of stone fruit, crown gall, oak root fungus, powdery mildew, rust
Strawberry	■ Aphids, cutworms, flea beetles, June beetles, leafrollers, mealybugs, nematodes, root weevils, scale insects, slugs and snails, spider mites, tarnished plant bugs, thrips, wireworms ■ Birds, deer, rabbits, squirrels ■ Anthracnose, gray mold, oak root fungus, powdery mildew, southern blight, verticillium wilt
Walnut	■ Aphids, borers, codling moths, curculios, fall webworms, fruit flies, lacebugs, leafrollers, mealybugs, scale insects, spider mites, tussock moths ■ Birds, deer, pocket gophers, squirrels ■ Crown gall, oak root fungus

FLOWERS

Aster	■ Aphids, blister beetles, cucumber beetles, European corn borers, leafhoppers, slugs and snails, tarnished plant bugs, thrips, whiteflies ■ Deer ■ Downy mildew, fusarium wilt, gray mold, mosaic viruses, powdery mildew, rust, verticillium wilt
Begonia	■ Aphids, mealybugs, nematodes, slugs and snails, spider mites, thrips, whiteflies ■ Pocket gophers, voles ■ Gray mold, powdery mildew
Carnation	■ Aphids, cabbage loopers, cutworms, mealybugs, slugs and snails, spider mites, thrips ■ Deer, squirrels ■ Bacterial wilt, gray mold, fusarium wilt, rust, southern blight
Chrysanthemum	■ Aphids, cabbage loopers, corn earworms, cucumber beetles, cutworms, European corn borers, lacebugs, leaf miners, nematodes, slugs and snails, spider mites, tarnished plant bugs, thrips, whiteflies ■ Squirrels ■ Crown gall, fusarium wilt, gray mold, mosaic viruses, powdery mildew, rust, southern blight, verticillium wilt
Dahlia	■ Aphids, cucumber beetles, European corn borers, leafhoppers, leaf miners, nematodes, slugs and snails, tarnished plant bugs ■ Deer, pocket gophers ■ Crown gall, fusarium wilt, gray mold, mosaic viruses, powdery mildew, rust, verticillium wilt
Delphinium	■ Aphids, cutworms, leafhoppers, slugs and snails, spider mites, thrips ■ Mosaic viruses, powdery mildew

Geranium	■ Aphids, geranium budworms, mealybugs, slugs and snails, spider mites, whiteflies ■ Deer ■ Gray mold, mosaic viruses, oak root fungus, rust, verticillium wilt
Gladiolus	■ Aphids, cutworms, European corn borers, spider mites, tarnished plant bugs, thrips ■ Pocket gophers, squirrels ■ Fusarium wilt, gray mold, mosaic viruses, southern blight
Hollyhock	■ Aphids, European corn borers, leafhoppers, mealybugs, nematodes, slugs and snails, spider mites, thrips, whiteflies ■ Deer ■ Powdery mildew, rust
Impatiens	■ Aphids, cucumber beetles, mealybugs, nematodes, scale insects, slugs and snails, spider mites, tarnished plant bugs ■ Rabbits ■ Southern blight
Lily	■ Aphids, mealybugs, nematodes, root maggots, scale insects, slugs and snails, spider mites, thrips ■ Pocket gophers, squirrels ■ Fusarium wilt, gray mold, mosaic viruses, southern blight
Marigold	■ Cabbage loopers, cutworms, leaf miners, slugs and snails, spider mites, tarnished plant bugs ■ Pocket gophers ■ Downy mildew, fusarium wilt, gray mold, mosaic viruses, southern blight, verticillium wilt
Nicotiana	■ Blister beetles, Colorado potato beetles, cutworms, geranium budworms, hornworms, slugs and snails ■ Mosaic viruses, powdery mildew
Pansy	■ Aphids, cutworms, flea beetles, mealybugs, nematodes, slugs and snails, spider mites, wireworms ■ Rabbits ■ Downy mildew, powdery mildew, southern wilt
Peony	■ Rose chafers, thrips ■ Fusarium wilt, gray mold, verticillium wilt
Petunia	■ Armyworms, blister beetles, Colorado potato beetles, flea beetles, geranium budworms, hornworms, slugs and snails ■ Gray mold, mosaic viruses, powdery mildew, verticillium wilt
Poppy	■ Aphids, corn earworms, leafhoppers, mealybugs, rose chafers, slugs and snails, tarnished plant bugs ■ Birds ■ Powdery mildew, verticillium wilt
Primrose	■ Leaf miners, slugs and snails, spider mites, whiteflies ■ Deer, rabbits ■ Gray mold
Snapdragon	■ Aphids, cabbage loopers, corn earworms, nematodes, slugs and snails, spider mites ■ Downy mildew, fusarium wilt, gray mold, mosaic viruses, powdery mildew, rust, verticillium wilt
Tulip	■ Aphids, spider mites, wireworms ■ Pocket gophers, rabbits, squirrels, voles ■ Fusarium wilt, mosaic viruses, southern blight
Zinnia	■ Aphids, blister beetles, cucumber beetles, European corn borers, flea beetles, Japanese beetles, mealybugs, tarnished plant bugs, thrips, whiteflies ■ Gray mold, mosaic viruses, powdery mildew

TREES & SHRUBS

Ash
■ Aphids, borers, lacebugs, scale insects, tent caterpillars, whiteflies   ■ Deer, voles
■ Anthracnose, verticillium wilt

Azalea & rhododendron
■ Aphids, borers, lacebugs, leaf miners, root weevils, scale insects, spider mites, thrips, whiteflies   ■ Pocket gophers, voles   ■ Powdery mildew, water mold root rot

Dogwood
■ Borers, cicadas, leafhoppers, scale insects   ■ Voles   ■ Anthracnose, powdery mildew

Elm
■ Aphids, bark beetles, borers, cankerworms, elm leaf beetles, leafhoppers, leafrollers, scale insects   ■ Deer, voles   ■ Anthracnose, Dutch elm disease, verticillium wilt

Euonymus
■ Borers, scale insects   ■ Deer, voles   ■ Crown gall, powdery mildew

Juniper
■ Aphids, bagworms, borers, leaf miners, scale insects, spider mites   ■ Oak root fungus, rust, water mold root rot

Lilac
■ Aphids, borers, cucumber beetles, leaf miners, mealybugs, scale insects, spider mites, whiteflies   ■ Powdery mildew, verticillium wilt

Maple
■ Aphids, bagworms, borers, cankerworms, gypsy moths, leafhoppers, mealybugs, nematodes, scale insects, spider mites, thrips, whiteflies   ■ Deer   ■ Anthracnose, powdery mildew, verticillium wilt

Oak
■ Aphids, bark beetles, borers, cankerworms, cicadas, gypsy moths, Japanese beetles, lacebugs, leaf miners, leafrollers, scale insects, spider mites, tent caterpillars, tussock moths, whiteflies   ■ Deer, pocket gophers, squirrels   ■ Anthracnose, oak root fungus, oak wilt, powdery mildew

Pine
■ Aphids, bark beetles, borers, nematodes, sawflies, scale insects, spider mites
■ Deer, pocket gophers   ■ Rust

Rose
■ Aphids, borers, corn earworms, cucumber beetles, curculios, fall webworms, harlequin bugs, Japanese beetles, leafhoppers, leafrollers, nematodes, rose chafers, sawflies, scale insects, slugs and snails, spider mites, thrips, whiteflies   ■ Deer, pocket gophers
■ Black spot, crown gall, downy mildew, gray mold, oak root fungus, powdery mildew, rust

Sycamore
■ Aphids, borers, lacebugs, psyllids, scale insects   ■ Anthracnose, powdery mildew

Willow
■ Aphids, borers, fall webworms, sawflies, scale insects, tussock moths   ■ Deer, rabbits
■ Rust

LAWNS

■ Armyworms, billbugs, chinch bugs, cutworms, grasshoppers, Japanese beetles, June beetles, rose chafers, sod webworms   ■ Dogs, moles, pocket gophers, raccoons, voles
■ Fairy rings, leaf spot, powdery mildew, pythium blight, rust, slime molds, snow molds, southern blight, tip burn

**NDEX**

Glossary

Cole crop A cruciferous plant, including cabbage, cauliflower, broccoli, Brussels sprouts, bok choy, collards, kale, kohlrabi, mustard, radish, rutabaga, and turnip.

Cucurbit A gourd-family crop, including cucumber, melon, pumpkin, and squash.

Honeydew Undigested plant sap excreted by sucking insects such as aphids, leafhoppers, mealybugs, psyllids, scale insects, and whiteflies.

Integrated Pest Management (IPM) An approach that focuses on preventing problems through good growing practices and by implementing physical, biological, and chemical controls only when needed.

Overwinter To survive the winter.

Pathogen A disease-causing agent, usually a fungus, bacterium, or virus.

Pesticide A substance that kills a particular type of organism. Different categories of pesticides (and the pests they target) include insecticides (insects), acaricides (mites), molluscicides (slugs and snails), rodenticides (rodents), fungicides (fungi), and herbicides (weeds).

Pheromone A chemical produced by an insect or other animal to attract other members of the same species.